Quasilinearization
and Nonlinear Problems
in Fluid *and* Orbital
Mechanics

Series (handwritten)

Modern Analytic *and* Computational Methods *in* Science *and* Mathematics

A GROUP OF MONOGRAPHS
AND ADVANCED TEXTBOOKS

Richard Bellman, EDITOR

University of Southern California

Published

1. R. E. Bellman, R. E. Kalaba, and Marcia C. Prestrud, Invariant Imbedding and Radiative Transfer in Slabs of Finite Thickness, 1963

2. R. E. Bellman, Harriet H. Kagiwada, R. E. Kalaba and Marcia C. Prestrud, Invariant Imbedding and Time-Dependent Transport Processes, 1964

3. R. E. Bellman and R. E. Kalaba, Quasilinearization and Nonlinear Boundary-Value Problems, 1965

4. R. E. Bellman, R. E. Kalaba, and Jo Ann Lockett, Numerical Inversion of the Laplace Transform: Applications to Biology, Economics, Engineering, and Physics, 1966

5. S. G. Mikhlin and K. L. Smolitskiy, Approximate Methods for Solution of Differential and Integral Equations, 1967

6. R. N. Adams and E. D. Denman, Wave Propagation and Turbulent Media, 1966

7. R. L. Stratonovich, Conditional Markov Processes and Their Application to the Theory of Optimal Control, 1968

8. A. G. Ivakhnenko and V. G. Lapa, Cybernetics and Forecasting Techniques, 1967

9. G. A. Chebotarev, Analytical and Numerical Methods of Celestial Mechanics, 1967

10. S. F. Feshchenko, N. I. Shkil', and L. D. Nikolenko, Asymptotic Methods in the Theory of Linear Differential Equations, 1967

11. A. G. Butkovskiy, Distributed Control Systems, 1969

12. R. E. Larson, State Increment Dynamic Programming, 1968

13. J. Kowalik and M. R. Osborne, Methods for Unconstrained Optimization Problems, 1968

14. S. J. Yakowitz, Mathematics of Adaptive Control Processes, 1969

15. S. K. Srinivasan, Stochastic Theory and Cascade Processes, 1969

16. D. U. von Rosenberg, Methods for the Numerical Solution of Partial Differential Equations, 1969

17. R. B. Banerji, Theory of Problems Solving: An Approach to Artificial Intelligence, 1969

18. R. Lattès and J.-L. Lions, The Method of Quasi-Reversibility: Applications to Partial Differential Equations. Translated from the French edition and edited by Richard Bellman, 1969

19. D. G. B. Edelen, Nonlocal Variations and Local Invariance of Fields, 1969

20. J. R. Radbill and G. A. McCue, Quasilinearization and Nonlinear Problems in Fluid and Orbital Mechanics, 1970

In Preparation

21. W. Squire, Integration for Engineers and Scientists

22. T. Pathasarathy and T. E. S. Raghavan, Some Topics in Two-Person Games

23. T. Hacker, Flight Stability and Control

24. D. H. Jacobson and D. Q. Mayne, Differential Dynamic Programming

25. H. Mine and S. Osaki, Markovian Decision Processes

26. W. Sierpiński, 250 Problems in Elementary Number Theory

27. E. D. Denman, Coupled Modes in Plasmas, Elastic Media, and Parametric Amplifiers

28. F. A. Northover, Applied Diffraction Theory

29. G. A. Phillipson, Identification of Distributed Systems

30. D. H. Moore, Heaviside Operational Calculus: An Elementary Foundation

Quasilinearization *and* Nonlinear Problems *in* Fluid *and* Orbital Mechanics

John R. Radbill

North American Rockwell Corporation,
Long Beach, California

and

Gary A. McCue

North American Rockwell Corporation,
Downey, California

American Elsevier
Publishing Company, Inc.
NEW YORK • 1970

AMERICAN ELSEVIER PUBLISHING COMPANY, INC.
52 Vanderbilt Avenue, New York, N.Y. 10007

ELSEVIER PUBLISHING COMPANY, LTD.
Barking, Essex, England

ELSEVIER PUBLISHING COMPANY
335 Jan Van Galenstraat, P.O. Box 211
Amsterdam, The Netherlands

Standard Book Number 444-00060-7

Library of Congress Card Number 69-11146

To the Spirit
of the
Space Sciences
Laboratory

Contents

Chapter 1

Linear Ordinary Differential Equations

Chapter 2

Quasilinearization

Chapter 3

Solution of Boundary Layer Equation

Chapter 4

Two Component Boundary Layers on an Ablating Wall

Chapter 5

Computation of Electrostatic
Probe Characteristics

Chapter 6

Prediction of the Stability of Laminar Boundary Layers

Chapter 7

Prediction of the Stability of Laminar Pipe Flow

Chapter 8

Optimum Orbital Transfer with "Bang-Bang" Control

Chapter 9

A Generalized Subroutine for Solving Quasilinearization Problems

Appendix 1

Preface

Throughout modern engineering and applied science, one is continually confronted with significant problems that will not yield to classical analytical methods of analysis. When such problems arise, the mathematician must turn his attention to the development of numerical methods and computational algorithms. Fortunately, today's electronic digital computers have expanded the application of numerical techniques. Using these devices, it is a straightforward matter to provide the numerical solution of several thousand simultaneous nonlinear differential equations subject to a complete set of initial conditions. This kind of computational speed and accuracy has led to radically new ideas and concepts for solving large systems of equations.

This book is concerned with the analysis, formulations, and computational sophistication involved in the use of quasilinearization, a modern mathematical technique, which is particularly suited to digital computer applications. It is a straightforward computational technique for solving systems of nonlinear differential equations, subject to two point or multiple point boundary conditions, in reasonable computing times. The method of quasilinearization leads to a scheme of successive approximations that converges rapidly (quadratically) and that requires moderate amounts of computation at each stage.

Computational techniques have been successfully applied to numerous problems in many diverse fields. Sometimes, as in celestial mechanics, orbit determination, and orbital transfer, the basic equations of the system are ordinary differential equations. In other areas, particularly in mathematical physics and mathematical biology, the underlying equations are partial differential equations, integral differential equations, differential difference equations, and equations of even more complex type. Extensive preliminary analyses of the precise problem formulation, and the analytical procedures to be used, are required in order to exploit fully the powerful computational capabilities now available. Whereas in earlier times the preferred procedure was to simplify so as to obtain linear functional equations, the current aim is to transform all computational questions to initial value problems for ordinary differential equations, whether linear or nonlinear. The theories of dynamic programming and invariant imbedding achieve this in a number of areas through the introduction of new state variables and the use of semigroup properties in space, time, and structure. Quasilinearization achieves this objective by combining linear approximation techniques with the capabilities of the digital computer. The approximations are carefully constructed to yield rapid convergence, and monotonicity as well, in many cases.

Quasilinearization was developed by Bellman and Kalaba from an origin in the theory of dynamic programming. The connections with classical geometry and analysis are, nonetheless, quite diverse and do not constitute any simple pattern. There is a strong heritage from the theory of differential inequalities started by Caplygin and continued by a very active Russian school. Intimately connected with this is the modern theory of positive and monotone operators, created by Collatz and vigorously developed by Bellman, Redheffer, and others. This, in turn, is associated with the theory of generalized convexity pursued by Beckenbach, Bonsall, Peixoto, and a number of other analysts. In many applications, the methods overlap the Newton-Raphson-Kantorovich approximation techniques in function space.

The objectives of the theory of quasilinearization are easily stated. First, we desire a uniform approach to the study of the existence and uniqueness of the solutions of ordinary and partial differential equations subject to initial- and boundary-value conditions. Second, we want representation theorems for these solutions in terms of the solutions of linear equations. Finally, we desire a uniform approach to the numerical solution of both descriptive and variational problems, which possesses various monotonicity properties and rapidity of convergence.

Goals

In writing this book our aim has been to explore the properties, behavior, and uses of quasilinearization. The material presented deals largely with practical engineering applications and is based upon experience gained during the solution of numerous significant problems in fluid and orbital mechanics. The problems treated range from the solution of a simple third-order differential equation to a system of ten nonlinear discontinuous Euler-Lagrange equations arising from an optimal control problem. Heavy emphasis necessarily has been placed upon numerical analysis, computational experience, and the tricks and ploys that are required for successful solution of realistic engineering problems. It is believed that this book represents a comprehensive treatment of actual applied science applications of quasilinearization that will enable readers to adapt the technique to their own particular problems.

In preparing this manuscript, it was assumed that the reader would have a working knowledge of applied mathematics equivalent to that of a college graduate with a degree in mathematics, engineering, or the physical sciences.

It was also assumed that the reader had at least a passing acquaintance with digital computers and numerical analysis. Although the latter knowledge is not required to understand most of the material in the chapters that follow, it would be very difficult to recreate the solutions presented without a good working knowledge of numerical computation on large digital computers; and some of the motivation for the procedures would be lost. The reader who is not a numerical analyst may wish to consult a text by such authors as Hildebrand, Kelly, or Ralston and Wilf. These references are but a few that will provide the necessary background on numerical integration, truncation errors, round-off errors, stability of integration processes, and the general aspects of numerical analysis.

Contents

By way of introduction let us turn our attention to a brief description of the contents of the nine chapters. To refresh the memory of the reader, Chapter 1 considers the properties and uses of linear ordinary differential equations. Throughout the discussions, available opportunities are utilized to introduce and define the notation and definitions of later chapters. By considering first-order linear differential equations, it is possible to introduce the concept of homogeneous and particular solutions and their relation to boundary conditions. It is also noted that the solution to a given differential equation need not be an explicit analytic expression, but may be a succession of different approximations on sequential intervals of the independent variable, or only tabulations. The discussion continues by considering second-order differential equations, multipoint boundary-value problems, and systems of linear ordinary differential equations.

The second chapter introduces the method of quasilinearization and indicates how it provides an extensive generalization of the Newton-Raphson method for solving nonlinear equations. During this discussion, its important quadratic convergence property is also explored. The multi-dimensional version is introduced and the basic properties of quasilinearization are explored in some detail. A concise statement of the quasilinearization method, which illustrates the application of the linear theory in Chapter 1, concludes Chapter 2.

Chapter 3 considers the first example of the application of quasilinearization. The problem considered is one involving fluid mechanics, where in it is desired to study laminar boundary layers on walls bordering flowing fluids. We examine the Falkner-Skan extension of the Blasius equation because it displays, in a simple equation, the nonlinear (bilinear) behavior typical of boundary-layer problems and two point boundary conditions.

Because the Falkner-Skan equation is comparatively simple, its solution does not require all of the capabilities of the general quasilinearization subroutines developed by the authors (these subroutines are discussed in Chapter 9). It does, however, make use of two point boundary conditions; i.e., some boundary conditions are specified at the initial point and others are specified at the opposite end of the interval. This example also illustrates the concise notation to be used in later chapters and provides a look at the kinds of convergence one may expect from typical quasilinearization applications when using an extremely crude initial approximation to the final solution.

The problem considered in Chapter 4 involves a two-component boundary layer on an ablating wall. The extensive preliminary analysis often required in applied science problems is illustrated by the use of a local similarity transformation on the original partial differential equations to produce a system of ordinary differential equations of total order seven. This application introduces the treatment of initial conditions that are related by algebraic equations. The quasilinearization process therefore involves subsidiary calculations. The equations, describing ablation, that interrelate the initial (wall) conditions were linearized in a manner similar to the linearization of the differential equations. The solution was obtained with the usual rapid convergence properties of quasilinearization.

Chapter 5 considers a problem of increased complexity, wherein one must utilize an asymptotic analytical solution at one end of a semi-infinite interval to obtain a finite numerical integration length. The problem involves the computation of continuum electrostatic probe characteristics and leads to a sixth-order nonlinear system of differential equations with two point boundary conditions and two parameters, which are determined as part of the solution. The asymptotic quasineutral solution is used as an initial condition for solution by quasilinearization between the edge of the plasma sheath and the probe surface. This additional complication is easily managed. This particular problem was also shown to have very rapidly growing solutions, which caused the homogeneous solution to become singular, thereby preventing solution. A Gramm-Schmidt orthogonalization process was required to overcome this difficulty. This particular orthogonalization technique has been incorporated into the authors' general quasilinearization program.

The prediction of stability of laminar boundary layers provides the subject matter of Chapter 6. The example considered is a characteristic value problem of large dimension. Quasilinearization was used to compute eigenfunctions and eigenvalues for the complete Orr-Summerfeld equation, equivalent to a fourth-order complex system. The problem again required asymptotic solutions for boundary conditions at one end of the interval, orthogonalization of the homogeneous solutions, and transformation to a new independent variable. The quasilinearization program calculated the real and imaginary parts of the eigenfunction, their first three derivatives, and the two corresponding eigenvalues.

A final fluid mechanics problem is presented in Chapter 7. The problem involves the stability of two-dimensional laminar channel flow. In order to apply quasilinearization effectively, it was necessary to introduce a "coordinate stretching" transformation to circumvent the destabilizing effect of a large parameter (the Reynolds number). The problem also illustrates the determination of higher characteristic functions (or modes) and the corresponding characteristic values. Convergence to a particular surface of the multiple valued solution is obtained by restricting the variation of the solution point to a particular region of parameter space during the initial iterations.

A final application of quasilinearization appears in Chapter 8. It presents an extremely sensitive optimum control problem and demonstrates the extreme care that must sometimes be exercised in order to achieve convergence. Quasilinearization was utilized to solve a discontinuous two point boundary-value problem that resulted from a variational formulation involving optimal orbital transfer. The boundary conditions were such that the transfer trajectory's end points could be assumed to be at unspecified positions upon arbitrary coplanar orbits. The vehicle was assumed to be thrust-limited and capable of controlling thrust direction and duration ("bang-bang" throttle control). It was found that accurate initial conditions, which were derived from the corresponding optimal impulsive orbital transfers, were required for convergence of the quasi-linearization process. An IBM 7094 double-precision computer program incorporating the above techniques then was utilized to generate optimal transfers between numerous pairs of arbitrary coplanar orbits. Double-precision was necessary in order to achieve convergence of this extremely sensitive problem. It was also necessary to employ several numerical "tricks" that were not required when solving the more stable fluid mechanics problems.

In order to present a comprehensive treatment of the subject, Chapter 9 and Appendix 1, respectively, contain a description and listing for the general quasilinearization program developed by the authors. The program is used to solve the very simple example in Chapter 3, which involves the laminar boundary layer on a wedge.

The applications we have examined are but a few that illustrate the effectiveness and potentialities of quasilinearization. It is our hope that others will be able to apply quasilinearization with successful results and that our experience will allow them to obtain solutions to increasingly more difficult problems in applied science and other fields.

Acknowledgements

Over the past few years we have received extensive help, encouragement, and inspiration from Doctors Richard Bellman and Robert Kalaba. We are particularly indebted to them for suggesting this monograph and for editing the final manuscript.

Much of the work would not have been accomplished had we not received continuing financial support. Several of the research studies upon which this book is based were funded by the North American Rockwell Corporation on Independent Research and Development funds. Portions of the astrodynamic research were supported by the National Aeronautics and Space Administration through the George C. Marshall Space Flight Center, under contracts NAS 8-5211, NAS 8-20238, and NAS 8-21077. Portions of the hydrodynamics research, which forms the basis of certain chapters, were supported by the United States Air Force -through the Office of Scientific Research of the Office of Aerospace Research, under contract AF 49(638)-1442.

We have imposed upon a number of friends and colleagues for their advice and comments. We gratefully acknowledge the assistance of Doctor J. E. McIntyre, whose frank suggestions materially improved the final manuscript. From the inception of this work, Mr. S. A. Jurovics and Doctor D. F. Bender offered numerous valuable contributions. We are also pleased to acknowledge the programming efforts of Mr. H. J. Duprie, the typing and manuscript preparation work of Mrs. C. K. Sinon and Miss L. Rubley, and the preparation of figures and illustrations, which was accomplished by Mrs. D. A. Jones.

In conclusion, we wish to recognize the inspiration and guidance provided by our many friends and colleagues who were associated with North American Rockwell Corporation's Space Sciences Laboratory. Notable among them was Paul R. Des Jardins, whose "neat but not gaudy" approach to programming permeates our work. Finally, and most important, we recognize the many tangible and intangible contributions of Doctor E. R. van Driest and Mr. H. W. Bell, who directed the Space Sciences Laboratory and who provided an environment wherein applied science flourished.

Long Beach, California
Downey, California
March, 1969

John R. RADBILL
Gary A. McCUE

LIST OF ILLUSTRATIONS

Chapter 1

LINEAR ORDINARY DIFFERENTIAL EQUATIONS

1.1 INITIAL VALUE PROBLEM

As a basis for the discussions of later chapters, it is important to review some elementary concepts of ordinary differential equations. This review will also serve to establish some of the notation that is employed herein. We start our discussion with the first-order linear ordinary differential equation. This simple type of equation serves to introduce the concept of homogeneous and particular solutions and their relation to boundary conditions. Using primes to denote differentiation by the independent variable t, we write

$$x' + a(t)x = f(t)$$
$$x(t_0) = c, \qquad t_0 \leqslant t \tag{1.1}$$

where $a(t)$ and $f(t)$ are arbitrary functions and t_0 is the initial value of the independent variable.

The solution of Eq. (1.1) is assumed to be composed of two parts: a particular solution $p(t)$, which reproduces the function $f(t)$ on the right-hand side when substituted into the left-hand side; and an homogeneous solution $h(t)$, which makes the left-hand side identically zero when substituted into it. This assumption is

$$x(t) = h(t) + p(t) \tag{1.2}$$

The homogeneous solution will be used to satisfy the initial boundary condition. Integrating the homogeneous equation [Eq. (1.1) with $f(t)$ replaced by zero], the homogeneous solution is found to be

$$h(t) = c_1 \exp\left[-\int_{t_0}^{t} a(t_1)\,dt_1\right] \tag{1.3}$$

where c_1 is a constant to be determined later and t_1 is a dummy variable.

In order to determine the particular solution, it is assumed to have the form

$$p(t) = C(t)\,h(t) \tag{1.4}$$

Substituting Eq. (1.3) into Eq. (1.4) and Eq. (1.4) into Eq. (1.1), we obtain after cancellation and integration,

$$C(t) = \int_{t_0}^{t} f(t_2) \exp\left[-\int_{t_0}^{t_2} a(t_1)dt_1\right] dt_2 -$$

and the particular solution becomes

$$p(t) = c_1 \exp\left[-\int_{t_0}^{t} a(t_1)dt_1\right] \int_{t_0}^{t} f(t_2) \exp\left[-\int_{t_0}^{t_2} a(t_1)dt_1\right] dt_2 \quad (1.5)$$

At $t = t_0$, if $a(t)$ and $f(t)$ are finite,

$$p(t_0) = 0, \qquad h(t_0) = c_1$$

Referring to Eqs. (1.1) and (1.2), we see that we must choose the constant c_1 in $h(t)$ so that

$$c_1 = c$$

Thus, the homogeneous solution is used to satisfy the boundary condition.

The general form of the first-order ordinary differential equation is represented by

$$x' = g(x, t)$$

$$x(t_0) = c \qquad (1.6)$$

where $g(x, t)$ is an arbitrary function that is likely to be nonlinear and may be given by a succession of different approximations on succeeding intervals of t or may be given only in tabular form. We shall encounter this last case in our mechanization of the quasilinearization algorithm in a computer program. This is discussed in chapter 9. An arbitrary function $g(x, t)$ of the type just described causes no concern because step-by-step numerical integration is easily mechanized on a digital computer. We assume that the reader has at least a passing acquaintance with digital computers and numerical analysis. If not, there are many good references on numerical integration that cover the problems of truncation errors, round-off errors, starting and changing step size, and stability [1–4]. We shall have further comments on the effects of errors in each chapter, and especially in connection with the very sensitive problem presented in Chapter 8. A tabulated function (at least if equally spaced on t) is ideally suited for application of numerical integration combined with interpolation.

In the numerical integration of Eq. (1.6), the solution $x(t)$ is started with the required initial condition $x(t_0) = c$, so the impossibility of superimposing solutions $h(t)$ and $p(t)$ causes no inconvenience. We shall see below that this loss of superposition in boundary-value problems requires radically different methods.

1.2   SECOND-ORDER DIFFERENTIAL EQUATIONS

We consider next the second-order linear ordinary differential equation of the form

$$y'' + a_1(t)y' + a_2(t)y = f(t) \tag{1.7}$$

with the initial boundary conditions

$$y'(t_0) = c_1, \qquad y(t_0) = c_2 \tag{1.8}$$

The approach here is again to look for a particular solution $p(t)$ to reproduce the right-hand side of Eq. (1.7) when substituted into the left-hand side. Two homogeneous solutions are now required, one for each of the two boundary conditions. These solutions are constructed to satisfy both the left-hand side of Eq. (1.7) identically and the initial conditions:

$$h_1(t_0) = 0, \qquad h_1'(t_0) = 1$$
$$h_2(t_0) = 1, \qquad h_2'(t_0) = 0 \tag{1.9}$$

Because of the linearity of Eq. (1.7), the particular and homogeneous solutions may be superimposed to give

$$y(t) = p(t) + [c_1 - p'(t_0)]h_1(t) + [c_2 - p(t_0)]h_2(t) \tag{1.10}$$

Although there are effective methods for obtaining analytic solutions for wide classes of second-order ordinary differential equations, and although analytic solutions are usually desired when obtainable, our concern here is with general numerical methods. In the case of a second-order linear equation, the term in y' may be eliminated by a transformation [1], so that the solution may be obtained by a double integration. This approach would be attractive from the standpoint of labor and integration accuracy. However, the transformation does not generalize to higher order equations and systems of equations; so we shall write Eq. (1.7) as two first-order equations by defining

$$X_1 = y', \qquad X_2 = y \tag{1.11}$$

The system of two equations equivalent to Eq. (1.7) then becomes

$$X_1' = -a_1(t)X_1 - a_2(t)X_2 + f(t)$$
$$X_2' = X_1 \tag{1.12}$$

with the boundary conditions

$$X_1(t_0) = c_1, \qquad X_2(t_0) = c_2 \tag{1.13}$$

Eqs. (1.12) can be integrated numerically step-by-step. The homogeneous solutions to Eqs. (1.12) [with $f(t)$ deleted] are started with the boundary conditions of Eq. (1.9), and a particular solution $p(t)$ is started with the initial conditions:

$$p(t_0) = p'(t_0) = 0 \qquad (1.14)$$

The solution to the initial-value problem is then given by Eq. (1.10). Of course, the initial conditions of Eq. (1.13) could have been used directly, and that approach would be valid even for a nonlinear second-order initial-value problem. However, in the next section, where the boundary conditions are not at the same point, the more elaborate apparatus of particular and homogeneous solutions will be required.

1.3   TWO POINT BOUNDARY-VALUE PROBLEMS

Suppose now that one boundary condition is given at each end of an interval $t_0 \leqslant t \leqslant t_1$, for the second-order linear ordinary differential equation of (1.7) or the system of (1.12). There are several possible combinations of boundary conditions, but let us examine the case

$$X_2(t_0) = b_1, \qquad X_2(t_1) = b_2 \qquad (1.15)$$

Again, Eq. (1.9) is used as the initial condition for the homogeneous solutions; and Eq. (1.14), as the initial condition for the particular solution. The particular and homogeneous solutions are integrated numerically from t_0 to t_1 to obtain some $p(t_1)$, which is generally not equal to b_2 and some values $h_1(t_1)$ and $h_2(t_1)$. In order to determine how much of the homogeneous solutions to add to the particular solution to satisfy the boundary conditions, we write the linear algebraic equations

$$A_1 h_1(t_0) + A_2 h_2(t_0) = b_1 - p(t_0)$$
$$A_1 h_1(t_1) + A_2 h_2(t_1 = b_2 - p(t_1) \qquad (1.16)$$

Upon solving Eq. (1.16) for A_1 and A_2, the solution that satisfies both boundary conditions is found to be

$$X_2 = A_1 h_1(t) + A_2 h_2(t) + p(t), \qquad X_1 = X_2' \qquad (1.17)$$

We note that this method of solution will not work if Eqs. (1.12) are nonlinear in X_1 and X_2 because the superposition of solutions leading to (1.16) is no longer valid. If a nonlinear system of equations with two point boundary values is first linearized in some manner, superposition can then be used to obtain a solution. The next chapter will show how this can be accomplished.

1.4 SYSTEMS OF LINEAR ORDINARY DIFFERENTIAL EQUATIONS

The progression from the system of two linear differential equations, associated with the second-order linear differential equation, to a system of arbitrary order is largely a matter of notation. The original set of differential equations may consist of a number of equations of different orders, e.g.,

$$\frac{d^m y}{dt^m} + a_1(t)\frac{d^{m-1} y}{dt^{m-1}} + \cdots + a_m(t)y = f(t)$$

$$\vdots \qquad\qquad\qquad\qquad\qquad (1.18)$$

$$\frac{d^n z}{dt^n} + p_1(t)\frac{d^{n-1} z}{dt^{n-1}} + \cdots + p_n(t)z = g(t)$$

where the equations are written in standard form, with the coefficient of the highest term made unity. By analogy with the treatment of the second-order equation, a new set of variables X_i are defined:

$$X_1 = d^{m-1}y/dt^{m-1}, \ldots X_m = y, \ldots$$

$$X_q = d^{n-1}z/dt^{n-1}, \ldots, \qquad X_{q+n} = z \qquad (1.19)$$

With the aid of these new variables, the system of equations is rewritten as a system of first-order differential equations in the form

$$X_1' = -a_1(t)X_1 + \cdots - a_m(t)X_m + f(t)$$

$$X_m' = X_{m-1}$$

$$\vdots \qquad\qquad\qquad\qquad\qquad (1.20)$$

$$X_q' = -p_1(t)X_q + \cdots - p_n(t)X_{q+n} + g(t)$$

$$\vdots$$

$$X_{q+n}' = X_{q+n-1}$$

where primes are used to denote differentiation by the independent variable t.

It is convenient to represent the right-hand sides of Eqs. (1.20) by a uniform notation, a set of functions $F_i(X_1, X_2, \ldots, X_{q+n}, t)$ defined by

$$F_1(X_1, X_2, \ldots, X_{q+n}, t) = -a_1(t)X_1 + \cdots - a_m(t)X_m + f(t) \quad (1.21)$$

$$F_2(X_1) = X_1, \text{ etc.}$$

A natural step at this point is to represent the subscripted variables and functions as elements of vectors. The vector composed of the variables is called the "state vector" and is defined by

$$\mathbf{X} = \{X_1, X_2, \ldots, X_n\} \tag{1.22}$$

for n dependent variables. The concept of a state vector comes from control theory and is a generalization of the thermodynamic concept. A state vector contains all of the dependent variables required to describe a mathematical model or system at one value of the independent variable (we limit ourselves to ordinary differential equations here). One or more parameters may also be elements of the state vector. These parameters are constants in the sense that they are not functions of the independent variable, but they are varied or solved as part of the solution of a problem. State vectors with parameters will be encountered in the electrostatic probe problem in Chapter 4 and in succeeding chapters.

A vector function $\mathbf{F}(\mathbf{X}, t)$ is defined in a manner similar to the state vector by

$$\mathbf{F}(\mathbf{X}, t) = \{F_1, F_2, \ldots, F_n\} \tag{1.23}$$

so that Eq. (1.20) becomes

$$\mathbf{X}' = \mathbf{F}(\mathbf{X}, t) \tag{1.24}$$

This very compact notation is not just an exercise in obscurantism. Besides minimizing the amount of writing required to describe the important features of a problem, the notation is an aid in programming a problem for a digital computer. The indexing of arrays is easily programmed in FORTRAN (and other languages of a similar sophistication) and is efficiently handled by the compiler and computer hardware.

Since we have assumed an nth-order system in Eqs. (1.22) and (1.23), we must satisfy n boundary conditions B_j, which may be at the beginning of the interval t_0, at the end t_f, or at intermediate locations. Although $\mathbf{F}(\mathbf{X}, t)$ is linear, in general, it will not be homogeneous; so a particular solution $\mathbf{P}(t)$ will be required to account for the inhomogeneous part as well as n homogeneous solutions $\mathbf{H}_j(t)$ to satisfy the boundary conditions. Here the particular and homogeneous solutions have been represented as vectors, in analogy with Eq. (1.22). In the same manner as was done for the second-order differential equation in Eq. (1.16), a set of linear algebraic equations is solved to determine the n combination coefficients A_j. We write

$$\sum_{j=1}^{n} A_j H_{mj}(t_i) = B_i - P_m(t_i) \quad [i = 1, 2, \ldots, n; \quad m = M(i); \quad t_i = L(i)] \tag{1.25}$$

where $M(i)$ specifies the component of the state vector by which the ith boundary condition is satisfied, and $L(i)$ is the location at which the ith boundary condition is specified. The doubly subscripted quantity H_{mj} represents the mth component of the jth homogeneous solution vector $\mathbf{H}_j$. The use of the function $M(i)$ is prompted by programming considerations because subscripted subscripts are not permitted in the FORTRAN programming language.

After having obtained the combination coefficients from Eq. (1.25), the desired solution $\mathbf{X}$ to Eq. (1.24), which satisfies all the boundary conditions, is found by summing the solutions in the form

$$\mathbf{X}(t) = \mathbf{P}(t) + \sum_{j=1}^{n} A_j \mathbf{H}_j(t) \tag{1.26}$$

In concluding this section, we observe that the homogeneous solution vectors must be independent (i.e., not parallel), or the matrix $[H_{mj}(t_i)]$ will be singular and the combination coefficients B_i will not be found in Eq. (1.25). We shall encounter this problem in Chapter 4, where an orthogonalization procedure is applied.

REFERENCES

1. F. B. Hildebrand, "Introduction to Numerical Analysis," p. 227. McGraw-Hill, New York, 1956.
2. L. G. Kelly, "Handbook of Numerical Methods and Applications," pp. 1–22. Addison-Wesley, Reading, Massachusetts, 1967.
3. W. E. Milne, "Numerical Solution of Differential Equations." Wiley, New York, 1953.
4. A. Ralston and H. S. Wilf, "Mathematical Methods for Digital Computers". Wiley, New York, 1959.

Chapter 2

QUASILINEARIZATION

2.1 INTRODUCTION

Before proceeding to the solution of problems in engineering and applied science by the quasilinearization method, it is important to gain an understanding of its basis and properties and some appreciation for its origin. We shall first trace one of the origins of the method in the determination of roots of nonlinear equations in algebra. Then we shall examine a second origin in the use of the maximum operation, as exemplified by the solution of the Riccati equation.

2.2 NEWTON-RAPHSON METHOD

One of the origins of quasilinearization is associated with the problem of finding a sequence of approximations to a root of the nonlinear scalar equation

$$f(x) = 0 \qquad (2.1)$$

We shall assume that $f(x)$ is monotone decreasing for all x in some sufficiently large neighborhood of the root and is strictly convex, i.e., $f''(x) > 0$. (See Fig. 2.1.) Hence, the root r is simple, provided that $f'(r) \neq 0$.

Let $x^{(0)}$ be an initial approximation to the root r, with $x^{(0)} < r\ [f(x^{(0)}) > 0]$, and suppose that we approximate to $f(x)$ by a linear function of x determined by the value and slope of the function $f(x)$ at $x = x^{(0)}$:[†]

$$f(x) \cong f(x^{(0)}) + (x - x^{(0)}) f'(x^{(0)}) \qquad (2.2)$$

Geometrically, this represents approximation of the curve by its tangent (see Fig. 2.2). A further approximation to r is then obtained by solving the linear equation in x:

$$f(x^{(0)}) + (x - x^{(0)}) f'(x^{(0)}) = 0 \qquad (2.3)$$

[†] Here and in future chapters the superscript in parenthesis is used to denote iteration number.

This yields the second approximation,

$$x^{(1)} = x^{(0)} - f(x^{(0)})/f'(x^{(0)}) \tag{2.4}$$

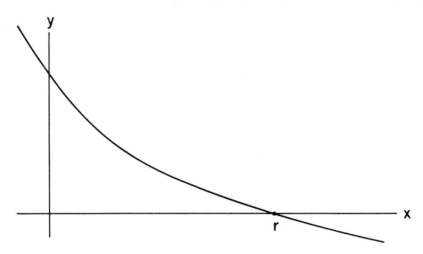

Figure 2.1.   The Root of a Convex Monotone Decreasing Function.

The process is then repeated at $x^{(1)}$, leading to a new value $x^{(2)}$, and at each new approximation, as illustrated in Fig. 2.2. The general recurrence relation is

$$x^{(n+1)} = x^{(n)} - f(x^{(n)})/f'(x^{(n)}) \tag{2.5}$$

It is clear from Fig. 2.2 that

$$x^{(0)} < x^{(1)} < x^{(2)} < \cdots < r \tag{2.6}$$

if $x^{(0)} < r$ initially. Analytically, this follows from the inequalities $f(x^{(n)}) > 0, f'(x^{(n)}) < 0$. This monotonicity is an important property computationally. We would prefer to have alternate approximations above and below r, i.e., $x^{(0)} < x^{(2)} < x^{(4)} < \cdots < r < \cdots < x^{(3)} < x^{(1)}$, but this is usually very difficult to arrange.

It is worthwhile to consider briefly the case where $x^{(0)} > r$ has been chosen. This might occur in practice if the function is costly to evaluate. A first evaluation of Eq. (2.5) will overshoot the root, but the new estimate falls in a region where $x^{(n)} < r$. Thus, monotone convergence would be obtained on the second and succeeding iterations. Of course, in real cases there is always the possibility that a sufficiently bad $x^{(0)}$ will produce an $x^{(1)}$ completely out of the neighborhood of the root.

A second property, even more important computationally, is perhaps not so obvious. It is called "quadratic convergence". We assert that

$$[x^{(n+1)} - r] \leqslant k[x^{(n)} - r]^2 \tag{2.7}$$

where k is independent of n, under a reasonable hypothesis concerning $f'''(x)$.

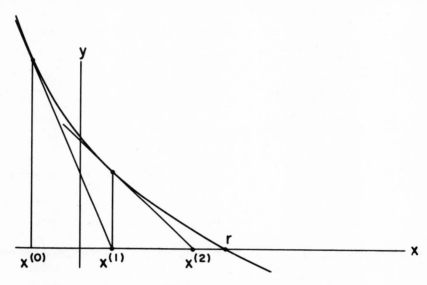

Figure 2.2.   A Convergent Sequence of Linear Approximations to a Root.

To see this, we write

$$x^{(n+1)} - r = x^{(n)} - \frac{f(x^{(n)})}{f'(x^{(n)})} - r$$

$$= x^{(n)} - \frac{f(x^{(n)})}{f'(x^{(n)})} - \left(r - \frac{f(r)}{f'(r)}\right)$$

$$= \phi(x^{(n)}) - \phi(r) \tag{2.8}$$

where $\phi(x) = x - f(x)/f'(x)$. Using the first three terms of the Taylor expansion, with a remainder, we obtain

$$x^{(n+1)} - r = (x^{(n)} - r)\phi'(r) + \frac{(x^{(n)} - r)^2}{2}\phi''(\theta) \tag{2.9}$$

where $x^{(0)} \leqslant x^{(n)} \leqslant \theta \leqslant r$. Since

$$\phi'(x) = [f(x) f''(x)/f'(x)^2] \tag{2.10}$$

we see that $\phi'(r) = 0$. Hence

$$|x^{(n+1)} - r| \leqslant k|x^{(n)} - r|^2 \tag{2.11}$$

where $k = \max_{x^{(0)} \leqslant \theta \leqslant r} \phi''(\theta)/2$, a bound depending upon $f'''(x)$. Furthermore,

$$x^{(n+1)} - x^{(n)} = \phi(x^{(n)}) - \phi(x^{(n-1)})$$

$$= (x^{(n)} - x^{(n-1)}) \phi'(x^{(n-1)}) + \frac{(x^{(n)} - x^{(n-1)})^2 \phi''(\theta)}{2} \tag{2.12}$$

where $x^{(n-1)} \leqslant \theta \leqslant x^{(n)}$, which leads to

$$(x^{(n+1)} - x^{(n)}) = (x^{(n)} - x^{(n-1)}) \frac{f(x^{(n-1)}) f''(x^{(n-1)})}{f'(x^{(n-1)})^2}$$

$$+ \frac{(x^{(n)} - x^{(n-1)})^2 \phi''(\theta)}{2} \tag{2.13}$$

Referring to Eq. (2.5), we replace $f(x^{(n-1)})/f'(x^{(n-1)})$ by $x^{(n)} - x^{(n-1)}$. Hence, Eq. (2.13) yields

$$(x^{(n+1)} - x^{(n)}) = (x^{(n)} - x^{(n-1)})^2 \left[\frac{f''(x^{(n-1)})}{f'(x^{(n-1)})} + \frac{\phi''(\theta)}{2} \right] \tag{2.14}$$

Thus

$$|x^{(n+1)} - x^{(n)}| \leqslant k_1 |x^{(n)} - x^{(n-1)}|^2 \tag{2.15}$$

where

$$k_1 = \max_{x_0 \leqslant \theta \leqslant r} \left[\frac{[f''(\theta)]}{[f'(\theta)]} + \frac{[\phi''(\theta)]}{2} \right] \tag{2.16}$$

The relation in Eq. (2.15) is also called "quadratic convergence".

It follows that as $x^{(n)}$ approaches r, there is an enormous acceleration of convergence. Asymptotically, each additional step doubles the number of correct digits in the approximation. In practice $f''(\theta)$ can be quite large or $f'(\theta)$ quite small, so that k_1 will be large, causing slow convergence during the early iterations; and the number of correct digits will not necessarily be doubled at each iteration. This is a most important property when large-scale operations are involved because the computing time is usually directly proportional to the number of iterations. The effect of round-off and truncation errors is merely to slow the rate of convergence, since each stage depends on fresh evaluations of the function and its derivative.

2.3 MULTIDIMENSIONAL VERSION

The method readily generalizes to higher dimensions. Consider the problem of determining a solution of a system of simultaneous equations

$$f_i(X_1, X_2, \ldots, X_n) = 0 \qquad (i = 1, 2, \ldots, N) \tag{2.17}$$

or, in vector notation,

$$\mathbf{f}(\mathbf{X}) = 0 \tag{2.18}$$

If $\mathbf{X}^{(0)}$ is an initial approximation, we write

$$\mathbf{f}(\mathbf{X}) = \mathbf{f}(\mathbf{X}^{(0)}) + J(\mathbf{X}^{(0)})(\mathbf{X} - \mathbf{X}^{(0)}) \tag{2.19}$$

where $J(\mathbf{X}^{(0)})$ is the Jacobian matrix:

$$J(\mathbf{X}^{(0)}) = [\partial f_i / \partial X_j]_{\mathbf{X} = \mathbf{X}^{(0)}} \tag{2.20}$$

The new approximation is thus

$$\mathbf{X}_1 = \mathbf{X}^{(0)} - J(\mathbf{X}^{(0)})^{-1} \mathbf{f}(\mathbf{X}^{(0)}) \tag{2.21}$$

and, generally, we obtain the recurrence relation:

$$\mathbf{X}^{(n)} = \mathbf{X}^{(n-1)} - J(\mathbf{X}^{(n-1)})^{-1} \mathbf{f}(\mathbf{X}^{(n-1)}) \tag{2.22}$$

It is not difficult now to obtain the analogs of the results of the previous section by appropriate assumptions concerning $\mathbf{f}(\mathbf{X})$ and its partial derivatives.

Analogous results hold for functional equations but require more sophisticated concepts both to state and to establish. See, e.g., Saaty [1].

2.4 GENERALIZATION OF THE NEWTON-RAPHSON METHOD TO ORDINARY DIFFERENTIAL EQUATIONS

One of the roots of quasilinearization lies in the application of the maximum operation [2] and the other lies in the Newton-Raphson method, which was generalized by Kantorovich and Krylov [3]. We wish to consider the latter line of development in the present section.

2.4.1 The First Order Equation

Let the first-order equation, Eq. (1.6), be represented as a transformation T on the dependent variable x, so that

$$y = T(x, t) = x' - g(x, t) \tag{2.23}$$

By analogy with the algebraic equation, we seek the "root" that makes the transformation $y = 0$ and that, in this case, corresponds to the solution $x(t)$. Let us imagine a three-dimensional space with coordinates t, $x(t)$, y. In each x, y plane, we wish to find the "root", i.e., the value of x that will make $y = 0$. Again by analogy with the Newton-Raphson derivation, y is expanded in the unknown function $x(t)$ about the approximation $x^{(0)}(t)$:

$$x' - g(x, t) = x'^{(0)} - g(x^{(0)}, t)$$

$$+ \frac{\partial x'^{(0)}}{\partial x}(x - x^{(0)}) - \frac{g(x^{(0)}, t)}{\partial x}(x - x^{(0)}) \tag{2.24}$$

where quadratic and higher terms in the Taylor series have been neglected. If $x(t)$ is the desired solution, the left-hand side will be zero. In like manner, the derivative x' can be expanded to give

$$x' = x'^{(0)} + (\partial x'^{(0)}/\partial x)(x - x^{(0)}) \tag{2.25}$$

When this is substituted into Eq. (2.24), we obtain

$$x' = g(x^{(0)}, t) + \frac{\partial g(x^{(0)}, t)}{\partial x}(x - x^{(0)}) \tag{2.26}$$

which is the quasilinearization algorithm for the first-order equation.

Of course, there is no two point boundary problem for a first-order differential equation, so quasilinearization is not required. However, the process can be visualized in 3-space for the first-order equation, whereas the geometric significance of second- and higher order equations is difficult to visualize.

2.4.2 The Second Order Equation

Although the visualization is somewhat more difficult, it is profitable to extend the derivation of the previous section to the second-order ordinary differential equation. We may write the second-order equation as two first-order equations, as in Section 1.3, but we now assume nonlinearity of right-hand-side functions $g_1(X_1, X_2)$ and $g_2(X_1, X_2)$:

$$X_1' = g_1(X_1, X_2), \qquad X_2' = g_2(X_1, X_2)$$

$$X_1(0) = b_1, \qquad X_2(t_1) = b_2$$

(2.27)

As in the previous section, we recognize that an arbitrary approximation will not satisfy the differential equations. This fact is expressed by writing

$$T_1(X_1, X_2, t) = X_1' - g_1(X_1, X_2)$$

$$T_2(X_1, X_2, t) = X_2' - g_2(X_1, X_2)$$

(2.28)

Now T_1 and T_2 may be plotted as surfaces, i.e., elevations above a plane with coordinates X_1, X_2, for each value of t. It is desired to find the intersection of the traces of the two surfaces on the zero plane if such an intersection (or intersections) exists. If the solution exists and is unique, there will be one and only one intersection.

We seek to find the "root" (X_1, X_2) of the functions in Eq. (2.28) by repetition of the following geometrical procedure: (1) construct a tangent plane to each surface at $(X_1^{(n)}, X_2^{(n)})$; (2) find the trace of these planes on the zero plane; (3) find the intersection of the two traces at $(X_1^{(n+1)}, X_2^{(n+1)})$ As the solution is approached, the approximation of the surfaces by planes becomes exact in the limit.

To formulate the analogous analytical procedure, we proceed as in the last section and expand Eq. (2.28) in Taylor series, keeping linear terms. The new values of the transformations $T_1(X_1^{(n+1)}, X_2^{(n+1)})$, $T_2(X_1^{(n+1)}, X_2^{(n+1)})$ are set equal to zero (corresponding to the intersection of the tangent planes with the zero plane) to yield the simultaneous equations:

$$X_1'^{(n)} - g_1(X_1^{(n)}, X_2^{(n)}) + \frac{\partial X_1'^{(n)}}{\partial X_1}(X_1^{(n+1)} - X_1^{(n)}) + \frac{\partial X_1'^{(n)}}{\partial X_2}(X_2^{(n+1)} - X_2^{(n)})$$

$$- \frac{\partial g_1}{\partial X_1}(X_1^{(n+1)} - X_1^{(n)}) - \frac{\partial g_1}{\partial X_2}(X_2^{(n+1)} - X_2^{(n)}) = 0$$

(2.29)

$$X_2'^{(n)} - g_2(X_1^{(n)}, X_2^{(n)}) + \frac{\partial X_2'^{(n)}}{\partial X_1}(X_1^{(n+1)} - X_1^{(n)}) + \frac{\partial X_2'^{(n)}}{\partial X_2}(X_2^{(n+1)} - X_2^{(n)})$$

$$\frac{\partial g_2}{\partial X_1'}(X_1^{(n+1)} - X_1^{(n)}) - \frac{\partial g_2}{\partial X_2'}(X_2^{(n+1)} - X_2^{(n)}) = 0$$

By expanding X_1 and X_2 in truncated Taylor series and substituting these into Eq. (2.29), the following new approximations to the derivatives are obtained:

$$X_1'^{(n+1)} = g_1(X_1^{(n)}, X_2^{(n)}) + \frac{\partial g_1}{\partial X_1}(X_1^{(n+1)} - X_1^{(n)}) + \frac{\partial g_1}{\partial X_2}(X_2^{(n+1)} - X_2^{(n)})$$

$$\text{(2.30)}$$

$$X_2'^{(n+1)} = g_2(X_1^{(n)}, X_2^{(n)}) + \frac{\partial g_2}{\partial X_1}(X_1^{(n+1)} - X_1^{(n)}) + \frac{\partial g_2}{\partial X_2}(X_2^{(n+1)} - X_2^{(n)})$$

This is seen to be the usual form of the quasilinearization algorithm applied to the second-order differential equation.

The geometric interpretation of quasilinearization continues to be valid when the number of elements in the state vector is extended to three or more. Here, however, we must contend with hypersurfaces and hyperplanes in hyperspace.

2.5 RICCATI EQUATION

Let us now apply this technique of quasilinearization to an important functional equation, the Riccati equation. The analysis that follows not only has tutorial value, but is useful in its own right because the Riccati equation, together with its multidimensional and function space analogs, plays a fundamental role in modern control theory and mathematical physics, particularly in connection with dynamic programming and invariant imbedding.

The Riccati equation is a first-order nonlinear differential equation having the form

$$v' + v^2 + p(t)v + q(t) = 0 \tag{2.31}$$

Despite its rather simple form, it cannot be solved explicitly in terms of quadratures and the elementary functions of analysis, for arbitrary coefficients $p(t)$ and $q(t)$.

The connection of the second-order linear differential equation with the Riccati equation is worth noting here. Starting with the second-order linear equation

$$u'' + p(t)u' + q(t)u = 0 \tag{2.32}$$

we employ the transformation

$$u = e^{\int v \, dt}, \qquad v = u'/u \tag{2.33}$$

in a purely formal manner and obtain (2.31) as the equation for v. We have reduced the order of the equation by one at the price of nonlinearity. As we have pointed out in Chapter 1, this is a small cost if numerical integration is to be used.

2.5.1 Solution of Riccati Equation in Terms of Maximum Operation

Let us now show that the Riccati equation can be solved in terms of a maximum operation. Consider the parabola $y = x^2$ and the tangent at a point $(x^{(1)}, y^{(1)}) = (x^{(1)}, x^{(1)^2})$,

$$y - y^{(1)} = 2x^{(1)}(x - x^{(1)}) \tag{2.34}$$

or

$$y = 2x^{(1)}x - x \tag{2.35}$$

(See Fig. 2.3).

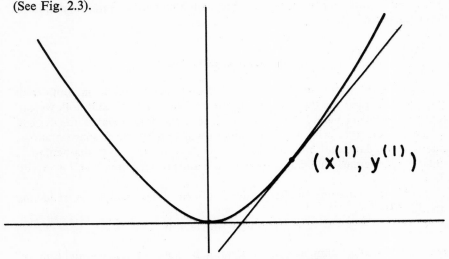

$$(x^{(1)}, y^{(1)})$$

Figure 2.3. An Example of the Maximum Operation.

Since the curve $y = x^2$, as a consequence of its convexity, lies above the tangent line, we have the inequality

$$x^2 \geqslant 2x^{(1)}x - x^{(1)^2} \tag{2.36}$$

for all $x^{(1)}$, with equality only if $x = x^{(1)}$. Hence

$$x^2 = \max_{x^{(1)}} [2x^{(1)}x - x^{(1)^2}] \tag{2.37}$$

A result that can, of course, be established immediately in many ways.

Returning to Eq. (2.31),

$$v' = -v^2 - p(t)v - q(t) \tag{2.38}$$

let us replace the quantity v^2 by its equivalent expression,

$$\max_u [2uv - u]^2$$

so that the equation now has the form

$$v' = -\max_u [2uv - u^2] - p(t)v - q(t)$$

$$= \min_u [u^2 - 2uv - p(t)v - q(t)] \tag{2.39}$$

Were it not for the minimization operation, the equation would be linear. As it is, the equation possesses certain properties ordinarily associated with linear equations. It is for these reasons that we employ the term "quasilinearization".

Consider next the related linear differential equation

$$w' = u^2 - 2uw - p(t)w - q(t) \tag{2.40}$$

where $u(t)$ is now a predetermined function of t. Let v and w have the same initial values $v(0) = w(0) = c$.

We suppose that the solution of the Riccati equation $v(t)$ exists in the interval $[0, t_0]$. The assumption of existence is necessary, since the solution of a Riccati equation need not exist for all t. Consider, as a counter example, the equation $u' = 1 + u^2$, $u(0) = 0$, with the solution $u = \tan t$ for $0 \leqslant t < \pi/2$, which does not exist at $t = \pi/2$.

To establish the inequality $w \geqslant v$, we argue as before. Observe that (2.39) is equivalent to the equation

$$v' \leqslant u^2 - 2uv - p(t)v - q(t) \tag{2.41}$$

for *all* $u(t)$, or

$$v' = u^2 - 2uv - p(t)v - q(t) - r(t) \tag{2.42}$$

where $r(t) \geqslant 0$, for $t \geqslant 0$. The function of $r(t)$ depends, of course, upon u and v, but we shall see below that its positivity is of primary importance.

Recall from Section 1.1 that the solution of the first-order linear differential equation can be expressed as a transformation on the coefficient and forcing functions. Thus, the solution of Eq. (2.42) may be written

$$v = T[-2u - p(t), \quad u^2 - q(t) - r(t)] \tag{2.43}$$

$$\leqslant T[-2u - p(t), \quad u^2 - q(t)] = w$$

where the positivity property of the operator $T(f, g)$ has been used to obtain the inequality.

Since this inequality holds for all $u(t)$, with equality for $u(t) = v(t)$, we have the desired result:

$$v(t) = \min_u T[-2u - p(t), \quad u^2 - q(t)] \tag{2.44}$$

which we can write explicitly as

$$v(t) = \min_u [c \exp\{-\int_0^t [2u(s) + p(s)] ds\}$$

$$+ \int_0^t \exp\{\int_s^t [2u(r) + p(r)] dr\} [u^2(s) - q(s)] ds] \tag{2.45}$$

We have sketched the relation of quasilinearization to maximum operation by application to the Riccati equation. Although the maximum operation will not be used directly in the examples of later chapters, this background will help to illuminate the properties of the method in the following sections.

2.5.2 Successive Approximations via Quasilinearization

We are now in a position to use the formula in Eq. (2.45) to generate a series of approximations to the solution $v(t)$. We know that the minimum is attained by the solution $v(t)$ itself. Hence, approximating the unknown function $r(t)$ in Eq. (2.42) by zero, we suspect that if v_0 is a reasonable initial approximation, then v_1, obtained as the solution of

$$v'^{(1)} = v^{(0)2} - 2v^{(0)}v^{(1)} - p(t)v^{(1)} - q(t)$$

$$v^{(1)}(0) = c_1 \tag{2.46}$$

will be even a better approximation by analogy with the procedure we employed to find the root of $f(x) = 0$, using the Newton-Raphson approximation. Repeating this approach, we obtain the recurrence relation

$$v'^{(n+1)} = v^{(n)2} - 2v^{(n)}v^{(n+1)} - p(t)v^{(n+1)} - q(t)$$

$$v^{(n+1)}(0) = c \tag{2.47}$$

This is precisely the recurrence relation we would obtain if we applied a Newton-Raphson-Kantorovich approximation scheme to the nonlinear differential equation

$$v' = -v^2 - p(t)v - q(t), \quad v(0) = c \tag{2.48}$$

Generally, if the differential equation were $v' = g(v, t)$ we would use the approximation scheme

$$v'^{(n+1)} = g(v^{(n)}, t) + (v^{(n+1)} - v^{(n)}) \frac{\partial g}{\partial v}(v^{(n)}, t)$$

(2.49)

$$v^{(n+1)}(0) = c$$

We wish to emphasize, however, that the two approaches are not equivalent despite the fact that they coincide in particular cases. As indicated by Bellman and Kalaba [2], quasilinearization owes a great deal to its inception within the theory of dynamic programming, with the concept of approximation in policy space playing an important and guiding role. In particular, from this viewpoint we are led to expect monotonicity of convergence. On the other hand, the geometric background of the Newton-Raphson approximation procedure leads us to expect quadratic convergence. As Bellman and Kalaba demonstrate [2], both properties are valid, and the resultant combination is extremely powerful computationally and analytically.

2.6 QUASILINEARIZATION ALGORITHM AND APPLICATION OF THE LINEAR THEORY

We conclude this chapter by summarizing the quasilinearization method in its general form before proceeding to particuiar applications in Chapters 3 through 8. The general, nonlinear multiple point boundary-value problem is concisely restated, followed by the general form of the quasilinearization algorithm. Finally, the solution indicated by the algorithm is obtained by application of the linear theory developed in Chapter 1.

The general, nonlinear multiple point boundary-value problem requires the solution of the nonlinear vector differential equation

$$\partial \mathbf{X}/\partial t = g(\mathbf{X}, t) \quad \text{where } \mathbf{X} = X_1, \ldots, X_n, \text{ and } \mathbf{g} = g_1(\mathbf{X}), \ldots, g_n(\mathbf{X}) \quad (2.50)$$

to be a nonlinear function, subject to the n boundary conditions:

$$X_{M_i}(t_i) = B_i \quad (i = 1, \ldots, n; \quad 1 \leqslant M_i \leqslant n; \quad t_0 \leqslant t_i \leqslant t_f) \quad (2.51)$$

Neither the M_i nor the values of the independent variable† t_i are necessarily all distinct, because the same component (variable) may satisfy boundary conditions at more than one point, and several different components may satisfy boundary conditions at the same point.

---

† e.g., time or distance.

Although the examples presented in Chapters 3 through 8 are for two point boundary problems ($t_i = t_{\text{initial}}$ or $t_f = t_{\text{final}}$), the method is valid for multiple boundary-value problems where one or more of the t_i lie inside the interval. A simple example of a multiple boundary-value problem is provided by a beam on three (or more) supports. For a more complex multiple point boundary-value problem involving determination of orbits from sighting angles, the reader is referred to Bellman and Kalaba [2].

The general form taken by the quasilinearization algorithm, as we shall use it in this book, whether derived from Newton-Raphson or the maximum operation [see Eq. (2.30) and Eq. (2.49)], is

$$\frac{dX_i^{(k+1)}}{dt} = g_i(\mathbf{X}^{(k)}, t) + \sum_{j=1}^{n} \frac{\partial g_i(X^{(k)}, t)}{\partial X_j} [X_j^{(k+1)} - X_j^{(k)}]$$

$$(i = 1, 2, \ldots, n; \qquad k = 0, 1, 2, \ldots) \qquad (2.52)$$

where the superscript in parenthesis denotes the iteration number.

Equation (2.52) is a linear differential equation in the $X_i^{(k+1)}$ since the $X_i^{(k)}$ are known functions of t obtained from the previous approximation. Thus, we are in a position to apply the theory of boundary-value problems, for linear ordinary differential equations, that was developed in Chapter 1.

Following Section 1.4, we obtain a particular solution $\mathbf{P}(t)$ of Eq. (2.52) and n homogeneous solutions $\mathbf{H}_j(t)$. For simplicity[†] the initial conditions for the particular solution are taken to be

$$\mathbf{P}(t_0) = 0 \qquad (2.53)$$

and the initial conditions for the $\mathbf{H}_j$ are taken to be

$$H_{ij}(t_0) = \delta_j^i \qquad \text{where} \quad \delta_j^i = \begin{matrix} 1, & i = j \\ 0, & i \neq j \end{matrix} \qquad (2.54)$$

In order to obtain a solution $\mathbf{X}^{(k+1)}$ of Eq. (2.52) that satisfies the boundary conditions of Eq. (2.51), we sum the particular and homogeneous solutions [see Eq. (1.25)] to obtain

$$\mathbf{X}^{(k+1)}(t) = \mathbf{P}(t) + \sum_{j=1}^{n} A_j \mathbf{H}_j(t) \qquad (2.55)$$

[†] We shall show in Chapter 3 that there is an advantage to using a less simple set of initial conditions.

where the combination coefficients A_j are obtained from Eq. (1.25), which we reproduce in the form

$$\sum_{j=1}^{n} A_j H_{mj}(t_i) = B_i - P_m(t_i) \qquad (i = 1, 2, \ldots, n; \qquad m = M_i) \qquad (2.56)$$

In practice $\mathbf{P}$ and the $\mathbf{H}_j$ are tabulated at a reasonably small number of points (depending on the higher derivatives of $\mathbf{X}^{(k)}$). Intermediate values of $\mathbf{X}^{(k+1)}$ are then obtained by interpolation. Experience has shown that $\mathbf{X}^{(k+1)}$ will be smoother than $\mathbf{P}$ or the $\mathbf{H}_j$, which is to be expected from the cancelling effects of the summation in Eq. (2.55).

The approximation $\mathbf{X}^{(k+1)}(t)$ is next fed back into the algorithm to obtain the approximation $\mathbf{X}^{(k+2)}(t)$, and the process is repeated until satisfactory convergence is secured. We have shown above that, with a sufficiently good initial approximation, the solution of Eq. (2.52) converges quadratically and monotonically (at least in the later stages) to the solution of Eq. (2.50). We shall see in Chapter 3 and the following chapters that, in some types of boundary-value problems, rapid convergence may be obtained even with crude approximations.

REFERENCES

1. T. L. Saaty, "Modern Nonlinear Equations", pp. 82–86. McGraw-Hill, New York, 1967.
2. R. E. Bellman and R. E. Kalaba, "Quasilinearization and Nonlinear Boundary-Value Problems", Elsevier, New York, 1965.
3. L. V. Kantorovich and V. I. Krylov, "Approximate Methods of Higher Analysis", Wiley (Interscience), New York, 1958.

Chapter 3

SOLUTION OF A BOUNDARY LAYER EQUATION

3.1 INTRODUCTION

As a first example of the application of quasilinearization to problems in fluid mechanics, we choose a problem that arises in the study of laminar boundary layers on walls bordering flowing fluids. We shall examine the Faulkner-Skan extension of the Blasius equation because it displays, in a simple equation, the nonlinear (bilinear) behavior typical of boundary-layer problems and the two point boundary conditions.

3.2 PHYSICAL BACKGROUND OF THE PROBLEM

When there is relative motion between a body of fluid and a solid boundary, the viscous effects are confined to a relatively thin layer adjacent to the boundary known as the "boundary layer" (Fig. 3.1). In this boundary layer the free-stream velocity drops rapidly to zero at the surface. Only here are the viscous stresses important, and the fluid outside this region can be treated as nonviscous or "perfect."

The flow of a viscous incompressible fluid is governed by the Navier-Stokes equations, which are nonlinear partial differential equations [1, 2]. Through an order-of-magnitude analysis due to Prandtl, the viscous stresses perpendicular to the boundary are dropped, thereby reducing the order of the equations from four to three. The resulting vector momentum equation (one component for each dimension) and scalar continuity equation may be further reduced to ordinary differential equations by use of a similarity transformation for certain simple geometries. That is, the velocity profiles at different distances along the boundary are similar.

For the case of flow over a two-dimensional wedge considered here, the similarity transformation uses a wall distance parameter η, defined by

$$\eta = y[(m+1)U/2vx]^{1/2} \tag{3.1}$$

where m is related to the total wedge angle $\pi\beta$ by

$$m = \beta/(2-\beta) \tag{3.2}$$

and $U(x) = u_1 x^m$ is the potential flow velocity. The similarity transformation

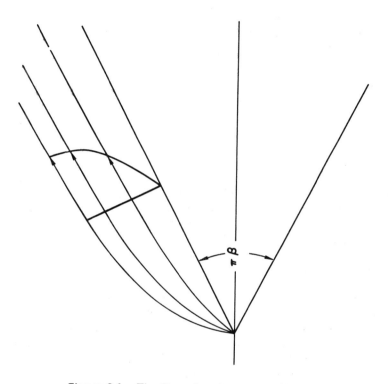

Figure 3.1. The Boundary Layer on a Wedge.

also requires the introduction of a stream function ψ, defined in terms of the x and y velocity components u and v by

$$u = \partial\psi/\partial y, \qquad v = -(\partial\psi/\partial x) \tag{3.3}$$

and assumed to be of the form

$$\psi(x, \eta) = [2vu_1/(m+1)]^{1/2} x^{(m+1)/2} f(\eta) \tag{3.4}$$

3.3 ORDINARY DIFFERENTIAL EQUATION

With the transformation [Eq. (3.4)], the partial differential equations governing the wedge flow reduce to the ordinary differential equation

$$f''' + ff'' + \beta(1 - f'^2) = 0 \tag{3.5}$$

where primes correspond to differentiation by η. At the wall,

$$\eta = 0: f = 0, \quad f' = 0 \tag{3.6a}$$

which correspond to both components of velocity being zero at the wall. At a great distance from the wall,

$$\eta \to \infty: f' \to 1 \tag{3.6b}$$

which corresponds to the approach to free-stream velocity.

In order to be strictly correct mathematically, we should examine the asymptotic behavior of Eq. (3.5) at large η. However, this would lead us too far afield at our present stage. The use of a limiting analytical solution as a boundary condition will be examined in Chapters 6 and 7. For the present, we take advantage of a well-known property of the Falkner-Skan equation: that the boundary condition, Eq. (3.6b), is approached rapidly beyond some finite η. That is, the boundary condition can be applied at some finite "edge" of the boundary layer with good accuracy, provided that $\beta = 0$, which corresponds to an accelerating flow.

As is typical of two point boundary-value problems, an initial boundary condition is missing, in this case f'', which is physically related to the shear stress at the wall. For a direct attack at integration of the equation, f'' must be estimated from physical experience or otherwise. In general the required boundary condition at η_{edge} is not satisfied. One method that has been used to find improved initial boundary conditions is to give several solutions with small changes in the approximate initial condition and to apply inverse interpolation (or extrapolation) onto the boundary values attained at the second point [3, 4]. A very real problem with this method is that the solution is quite sensitive to the approximated boundary condition and may grow so rapidly as to cause exponent overflows in the computer long before the second point is approached. This problem becomes rapidly more serious as the number of unknown initial conditions increases. The method of inverse interpolation, when applied to this equation, is susceptible to instabilities at large values of η as η_{edge} is approached. There may also be large changes in the solution at interior points for small errors in the boundary conditions.

Picard's method (successive substitutions) [5] has been used effectively on boundary-layer equations. However, the convergence is slow if the initial guess is not sufficiently good, and is only linear at best.

3.4 APPLICATION OF GENERAL METHOD

Let us now establish the correspondence between the general description of the method in terms of state variables in Chapter 1 and the simple nonlinear equation (3.5). The first step is to rewrite Eq. (3.4) as three first-order differential equations in the form

$$
\begin{aligned}
(f'')' &= -f(f'') - \beta[1 - (f')^2] \\
(f')' &= (f'') \\
(f)' &= (f')
\end{aligned}
\tag{3.7}
$$

The state vector $\mathbf{X}$ is seen to be

$$
\mathbf{X} = \{f'', f', f\}
\tag{3.8}
$$

and the vector function $\mathbf{g}$, the right-hand side of the equation, is

$$
\mathbf{g} = \{-ff'' - \beta[1 - f'^2], f'', f'\}
\tag{3.9}
$$

Carrying out the simple differentiation, the Jacobian matrix $[\partial g_i/\partial X_j]$ is found to be

$$
\left[\frac{\partial g_i}{\partial X_j}\right] =
\begin{bmatrix}
-f & 2f' & -f'' \\
1 & 0 & 0 \\
0 & 1 & 0
\end{bmatrix}
\tag{3.10}
$$

Boundary conditions are specified, for Eq. (3.7), in the form of three vectors, which facilitate retrieval of this information during the computer solution. The components of the first vector $\mathbf{B}$ specify the values of the n boundary conditions, where n is the order of the differential equation [e.g., 3 for Eq. (3.4)]. Components of the second vector $\mathbf{L}$ specify the locations at which the values of $\mathbf{B}$ must be satisfied. In the general case, the n boundary conditions can be specified on the same component of the state vector or on an arbitrary combination of components. This specification is provided by the vector $\mathbf{M}$. Thus, the component M_i must have the value B_i at the location (value of independent variable) L_i.

In terms of the current example, these vectors become

$$
\mathbf{B} = \{0, 0, 1\}, \quad \mathbf{L} = \{0, 0, \eta_{\text{edge}}\}, \quad \mathbf{M} = \{2, 3, 2\}
\tag{3.11}
$$

The old approximation $\mathbf{X}^{(k)}(\eta)$ must be generated by integration or stored for use in the quasilinearization algorithm, as has been pointed out in Chapter 1. In this example, and in the following chapters, we shall assume that sufficient computer storage is available for $\mathbf{X}^{(k)}(\eta)$, provided that we use a coarse spacing and obtain intermediate values by interpolation. The values of the independent variable (e.g., η) at which $\mathbf{X}^{(k)}(\eta)$ is stored then comprise an additional one-dimensional array, V_{ind}, and the components of $\mathbf{L}$ become integers referring to the location in V_{ind}.

3.5 ALTERNATIVE TREATMENT OF INITIAL CONDITIONS

No distinction has been made in the previous section between initial boundary conditions and conditions imposed at the end or interior of the interval of the independent variable. The present example has two initial conditions and one final condition. Referring to the discussion of the solution of linear ordinary differential equations in Section 1.4 and to the discussion of the quasilinearization algorithm, we recall that a solution to a particular problem can be obtained by determining a particular solution $\mathbf{P}(\eta)$ and n homogeneous solutions $\mathbf{H}_j(\eta)$ to satisfy the n boundary conditions. Proceeding as outlined in Section 3.2: The initial conditions on the particular solution may be taken as all zero and the initial condition for each of the $n = 3$ homogeneous solutions may be taken as equal to zero, except for a different nonzero element in each solution. Thus

$$\mathbf{P}(0) = \{0, 0, 0\}, \qquad [H_{ij}(0)] = \begin{bmatrix} 1 & 0 & 0 \\ 0 & 1 & 0 \\ 0 & 0 & 1 \end{bmatrix} \qquad (3.12)$$

Although this crude approximation to the solution works with this simple example, it will not work where convergence is very sensitive to the initial approximation. This is true of the bounded control problem in Chapter 8. In such cases, the particular solution must carry the maximum amount of significance. That is, the particular solution must have initial conditions consistent with the best available estimate $\mathbf{X}^{(k)}(\eta)$.

The homogeneous solutions should be small perturbations or corrections to the particular solution. Elements on the diagonal of $[H_{ij}(\eta)]$ may have to be small numbers to avoid the exponent overflow problem arising in machine computation, mentioned in connection with $\mathbf{P}(\eta)$. A method of scaling the homogeneous solutions to avoid overflow will be discussed in the next chapter.

Since the particular solution is now to be started with the best available initial conditions, the required initial conditions will be automatically satisfied. Only conditions that are at intermediate or final points remain to be considered. Thus, the homogeneous solutions corresponding to the initial conditions can be eliminated. With these considerations, Eq. (3.12) becomes

$$\mathbf{P}(0) = \{f_0'', 0, 0\}, \qquad \mathbf{H}_1(0) = \{\varepsilon, 0, 0\}$$

where f_0'' is an initial estimate or the value determined by the previous iteration, and ε is about 1% of the initial guess. By requiring that the particular solution satisfy the initial conditions, the number of differential equations to be integrated has been reduced by a factor of two in this example. The savings can be quite large when the order of the differential equation is large.

3.6 COMPUTER SOLUTION

Listings of a computer program for the present example are given in Appendix 1 and are part of an explanation of the use of a general quasi-linearization subroutine. The Faulkner-Skan equation does not require all of the features of the general subroutine presented in Appendix 1, but it does serve to illustrate some of the more important facets of its use.

The example was originally programmed in FORTAN II and the code used was a much simpler "ancestor" of that shown in the appendix. The former results are shown in Fig. 3.2. That solution was carried out on an IBM 7094 computer using Sterling interpolation and Adams-Moulton integration. The solution was stored at 26 values of the independent variable η. Additional details of the numerical solution are given in the original note [6]. What we shall examine here is the convergence behavior typical of the quasilinearization method.

The first (initial), second, third, and fifth (final) profiles of f'', f', and f are shown in Fig. 3.2. The initial approximation (solid lines) has intentionally been chosen far from the solution to illustrate the convergence of the method. Certainly, a much closer initial approximation than the crude one used here should be easy to obtain, for most boundary-layer problems. However, in simple problems the expenditure of a large effort on initial approximations is not worthwhile since the convergence is rapid.

The second approximation overshoots the final curve, and succeeding approximations converge rapidly from that side. The overshoot is not unexpected because the quasilinearization algorithm is equivalent to

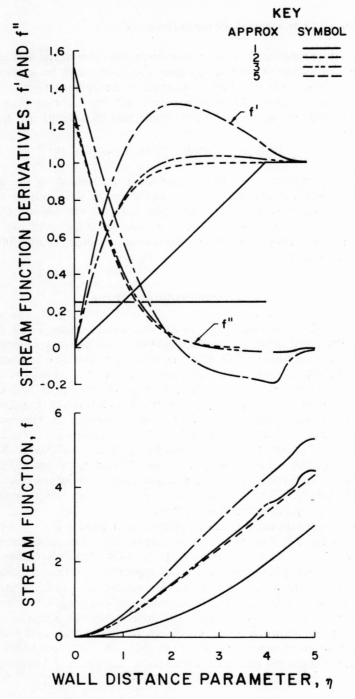

Figure 3.2. Convergence of the Solution to the
Faulkner-Skan Equation.

28

taking a tangent plane to a curved surface in function space. Note that the largest proportional deviations in the second approximation are found in the center of the f' (velocity) profile, since both ends are forced to satisfy the boundary conditions by the process of adding solutions. The strange bumps in $f^{(3)}$ are produced by the large curvatures in $f'^{(2)}$ and $f''^{(2)}$, but they have disappeared in $f^{(5)}$. The behavior of these curves is typical of convergence during quasilinearization solution of boundary-layer equations. In more complicated problems, such as those discussed in later chapters, some restriction on the initial changes of the approximations may be required to secure convergence.

Values of all functions in the fifth approximation differ from the corresponding functions in the fourth approximation at each of the 26 tabular (storage) points by less than 5 in the third decimal place. The value of f'' (shear stress) in the fifth approximation was 1.2397 compared with the published value [7] of 1.2326. This accuracy is consistent with the truncation error of 0.001 specification to the integration routine.

No instability is shown by the final profiles, although the velocity f' approaches its asymptotic value of unity at about 2.5, and the boundary condition was imposed at an η_{edge} of 5. The inverse interpolation method, mentioned at the beginning of the chapter, is very sensitive to instability in this region.

REFERENCES

1. S. Goldstein, "Modern Development in Fluid Dynamics," p. 90. Dover, New York, 1965.
2. H. Schlichting, "Boundary Layer Theory," p. 42. McGraw-Hill, New York, 1960.
3. J. Radbill, "Laminar Boundary Layer Computer Programs," North Am. Aviation, Space Inform. Systems Div. Rep. SID 64-1039.
4. W. E. Milne, "Numerical Solution of Differential Equations," p. 102. Wiley, New York, 1953.
5. W. E. Milne, *Ibid*, p. 106.
6. J. Radbill, "Application of quasilinearization to boundary layer equations," *AIAA J.*, 1, 1860 (1964).
7. H. Schlichting, "Boundary Layer Theory," p. 121. McGraw-Hill, New York, 1960.

Fn is redundant, see Section 2.1.

Chapter 4

TWO COMPONENT BOUNDARY LAYERS ON AN ABLATING WALL

4.1 INTRODUCTION

In this chapter we shall examine the solution of a seventh-order system of equations arising from a problem involving a boundary layer on an ablating wall. Of even greater interest than the increase of the order over that of the problem in Chapter 3, is the form of the initial boundary condition. In addition to known or unknown constant values, we are confronted with initial values of three variables related by two equations. The differential equations themselves display a greater nonlinearity, because of the variable fluid properties, than did the system of the previous chapter. Another notable feature is the use of a transformation to put the equations in a form for which a good initial estimate can easily be made.

4.2 PHYSICAL BACKGROUND OF THE PROBLEM

When gas, having a sufficiently high temperature, flows past a solid wall, the wall will ablate. By ablation we signify that the material of the wall will melt, vaporize, or decompose and be carried away by the stream of gas. The wall material may or may not react with the gas, but a large amount of energy is absorbed by the wall material in the process, so that the flow of heat to the lower layers of the wall is greatly reduced. The very large amount of energy that is absorbed by an ablative coating and the accompanying blocking of heat transfer to the interior of the reentry body makes ballistic missile warheads and manned space vehicles possible.

In this high-speed, compressible boundary-layer problem, the motion is no longer decoupled from the thermal energy as it was in the incompressible boundary layer of the last chapter. As a result, an energy equation of second order must be included with the equation of motion. In addition a continuity equation must be added to describe the diffusion of each species of ablating material or reactant (less one species because of overall continuity).

For realistic ablators, at high Mach numbers, the reaction kinetics of the chemical processes in the boundary layer completely dominate the dynamics by their complexity and nonlinearity. However, we shall greatly simplify the problem in order to display some important mathematical features and to avoid using an inordinate amount of space. Thus, all of the heat entering the wall is assumed to cause vaporization of material that is assumed to yield a single nonreacting diffusing species. Furthermore, all of the ablation is assumed to take place at a fixed ablation temperature requiring a constant ablation enthalpy per unit mass.

The equations given in the next section are in dimensionless form. Lengths are divided by a characteristic length L; all other variables are non-dimensionalized with respect to their values at the edge of the boundary layer. As a result of this procedure, the four dimensionless groups governing the problem are also evaluated at the edge of the boundary layer. These dimensionless groups are: the Reynolds number Re, the Mach number M, the Prandtl number Pr, and the Schmidt or Lewis numbers Sc or Le.

4.2.1 Basic Equations

The well-known equations for a steady two-component laminar boundary layer in two dimensions (i.e., on a surface with a large radius of curvature) are [1]:

Overall continuity:

$$\partial u/\partial x + \partial v/\partial y = 0 \tag{4.1}$$

Motion:

$$u\frac{\partial u}{\partial x} + v\frac{\partial u}{\partial y} = \frac{1}{\rho \text{Re}}\frac{\partial}{\partial y}\left\{\mu\frac{\partial u}{\partial y}\right\} + \frac{\partial U}{\partial x} \tag{4.2}$$

Energy:

$$\rho\left\{u\frac{\partial h}{\partial x} + v\frac{\partial h}{\partial y}\right\} = \frac{1}{\text{Pr Re}}\frac{\partial}{\partial y}\left\{k\frac{\partial T}{\partial y} + (h_1 - h_2)\frac{\rho D}{\text{Sc}}\frac{\partial c}{\partial y}\right\}$$

$$+ \frac{\text{M}^2}{\text{Re}}(\gamma - 1)\mu\left(\frac{\partial u}{\partial y}\right)^2 + \frac{\text{M}^2}{\text{Pr}}(\gamma - 1)u\frac{\partial U}{\partial x} \tag{4.3}$$

Single species continuity:

$$\rho\left\{u\frac{\partial c}{\partial x} + v\frac{\partial c}{\partial y}\right\} = \frac{1}{\text{Re Sc}}\frac{\partial}{\partial y}\left\{\rho D\frac{\partial c}{\partial y}\right\} \tag{4.4}$$

The notation, defined in the nomenclature section at the end of the chapter, is almost entirely conventional. Since only the concentration of the ablated material c appears, it is not subscripted. Likewise, the binary diffusion coefficient of the ablated material, with respect to air, is simply denoted by D.

In general both the dimensionless thermodynamic properties h, h_1, h_2, and ρ, as well as the dimensionless transport properties μ, k, and D, are functions of concentration. It is assumed that μ and k have the same temperature and concentration dependence, i.e., that the dimensionless quantities are equal. Since the pressure is assumed constant across the boundary layer, in conventional boundary-layer theory, and properties have been nondimensionalized with respect to their values at the edge of the boundary layer, the pressure dependence will not be considered explicitly.

In the example shown in the latter part of this chapter, the properties were further simplified to reduce the amount of computation, while examining the effect of the ablation boundary condition. The thermodynamic properties were assumed to follow the Gibbs-Dalton law with constant specific heats for the individual components (i.e., ablatant and air). Prandtl number, Lewis number, and the dimensionless transport properties were taken to be constant.

The boundary conditions placed on Eqs. (4.1) through (4.4) are of two types. The first type, which are the usual constant conditions specified at the wall and asymptotically at the edge of the boundary layer, are

$$
\begin{aligned}
y = 0 &: u = 0, \quad T = T_w \\
y \to \infty &: u = 1, \quad T = T_e, \quad c = c_e
\end{aligned}
\tag{4.5}
$$

The second type of condition specifies a relation between the values of variables on the boundary without specifying the values. In this case we have two such relations, which are

$$
\rho v \, \mathrm{Re} = \frac{-\rho D}{\mathrm{Sc}(1-c)} \frac{\partial c}{\partial y} = \frac{k}{\mathrm{Pr}(h_1 - h_{1s})} \frac{\partial T}{\partial y}
\tag{4.6}
$$

These relate the mass flux normal to the surface to the concentration and temperature gradients. As was noted above, all the heat is assumed to be used in vaporizing material with an enthalpy of vaporization of $h_1 - h_{1s}$, and none is conducted into the interior of the solid. Equations (4.6) constitute two relations among the four unknowns v, c, $\partial c/\partial y$, $\partial T/\partial y$. We vary two of these in the process of matching the boundary conditions at the edge of the boundary layer and determine the other two conditions, by means of Eq. (4.6), from those that are varied.

4.2.2 Dorodnitsyn Transformation

We now introduce a coordinate transformation coupled with an assumption for a stream function ψ that reduces the partial differential equations to ordinary differential equations. The equation of motion has a form that is nearly the same as that found in Chapter 3. This allows use of solutions to the incompressible boundary layer as initial approximations for the stream function and its derivatives. The transformation is

$$\chi = \text{Re}^{-1} \int_o^x u_e(x_1)\,dx_1, \qquad \eta = (2\chi)^{-1/2} \int_o^y [\rho(y_1)/\rho_e]\,dy_1 \qquad (4.7)$$

where the subscript e denotes a quantity at the edge of the boundary layer. A locally similar solution is obtained by introducing the stream function:

$$\psi = (2\chi)^{1/2} f(\eta) \qquad (4.8)$$

and by assuming that the velocity at the edge of the boundary layer can be approximated by

$$u_e(x) = x^m \qquad (4.9)$$

The stagnation enthalpy h_o is given by

$$h_o = h + M^2(\gamma - 1)\, u^2/2 \qquad (4.10)$$

and a stagnation enthalpy ratio is given by

$$H(\eta) = (h_o - h_{ow})/(h_{oe} - h_{ow}) \qquad (4.11)$$

where subscript w denotes the wall at $\eta = 0$.

4.2.3 Transformed Equations

After a considerable amount of manipulation, the following set of equations is obtained:

Motion:

$$\tau' + ff'' + \beta(\rho_e/\rho - f'^2) = 0 \qquad (4.12)$$

Energy:

$$Q' + fH' = 0 \qquad (4.13)$$

Single species continuity:

$$J' + fc' = 0 \qquad (4.14)$$

where

$$\tau = \rho\mu f'' \tag{4.15}$$

$$Q = \frac{\rho\mu}{Pr}\{H' + (h_{oe} - h_{ow})^{-1}[A(Pr-1)f'f'' + (Le-1)c + (h_1 - h_2)J]\} \tag{4.16}$$

$$J = c'\rho\mu\,Le/Pr \tag{4.17}$$

$$A = (\gamma-1)M^2 u_e^2, \qquad Le = Pr/Sc = \text{Lewis Number}$$

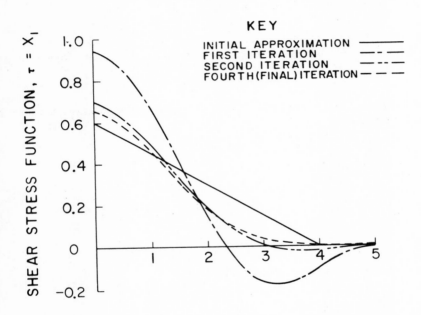

KEY

INITIAL APPROXIMATION ————
FIRST ITERATION —— · ——
SECOND ITERATION —— ·· ——
FOURTH(FINAL)ITERATION —— —— ——

SHEAR STRESS FUNCTION, $\tau = X_1$

DORODNITSYN WALL DISTANCE PARAMETER, η

Figure 4.1. Quasilinearization Solution for the
Boundary Layer on an Ablating Wall, Convergence of
Approximations for the Shear-Stress Function, τ.

Note that the second term in the braces in Eq. (4.16) will become small as the Prandtl and Lewis numbers approach unity, which is nearly true in practice.

The ablation conditions at the wall, Eq. (4.6), transform to

$$\frac{J}{1-c} = f = -\frac{Q(h_{ow}-\bar{h}_{oe})}{h_1-h_{1s}-(h_1-h_2)(1-c)} \tag{4.18}$$

and the other boundary conditions become

$$\eta = 0: f' = 0, \; H = 0; \qquad \eta \to \infty: f' \to 1, \; H \to 1, \; c \to 0$$

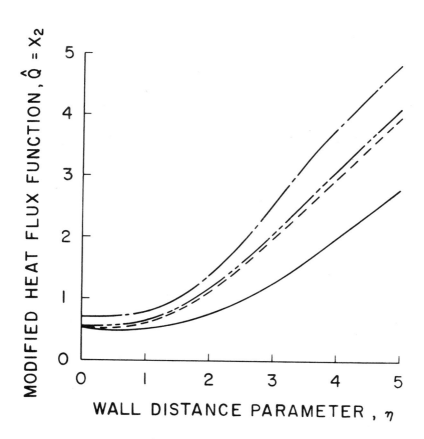

Figure 4.2. Quasilinearization Solution for the
Boundary Layer on an Ablating Wall, Convergence of
Approximations for the Modified Heat Flux Function, Q.

4.2.4 Additional Transformation of Heat and Mass Fluxes

As a preliminary to applying the quasilinearization algorithm, we integrate the energy equation (4.13) and the single-species continuity equation (4.14) by parts, and define modified heat- and mass-flux vectors by

$$\hat{Q}' = (Q+fH)' = f'H$$
$$\hat{J}' = (J+fc)' = f'c \qquad (4.19)$$

This transformation decreases the number of nonzero elements in the Jacobian matrix, which is shown below, but it also increases the complication of the remaining elements.

4.3 APPLICATION OF QUASILINEARIZATION

We are now prepared to apply the quasilinearization method to the transformed equations. We shall first establish the equivalence between the general form and the particular problem, and then we shall examine modifications that are necessary for the new type of boundary conditions.

4.3.1 State Vector and Jacobian Matrix

The state vector $\mathbf{X}$, discussed in Section 2.4, is defined in this problem to be

$$\mathbf{X} = \{\tau, \hat{Q}, c, f', H, \hat{J}, f\} \qquad (4.20)$$

This order of the components was originally chosen because the logic of the version of the computer program at that time required that the variables with unknown initial conditions be placed first. This restriction has been removed in the version of the computer program discussed in Chapter 9, but we shall maintain the original order of the vector here. The components of the vector function $\mathbf{g}(\mathbf{X})$ are

$$g_1 = -(f\tau/\rho\mu) - \beta\{(\rho_e/\rho) - f'^2\}$$

$$g_2 = f'H, \qquad g_3 = J\mathrm{Pr}/\mathrm{Le}\,\rho\mu, \qquad g_4 = \tau/\rho\mu \qquad (4.21)$$

$$g_5 = \frac{\mathrm{Pr}}{\rho\mu}\left\{\hat{Q} - fH - \frac{1}{h_{oe} - h_{ow}}\left[A\frac{(\mathrm{Pr}-1)}{\mathrm{Pr}}f' + (J-fc)\frac{(\mathrm{Le}-1)}{\mathrm{Le}}(h_1 - h_2)\right]\right\}$$

$$g_6 = f'c, \qquad g_7 = f'$$

The Jacobian matrix $[J_{ij}]$ has 49 elements, many of which require complicated expressions. As a result, we shall give the nonzero elements in order:

$$J_{11} = \frac{-f}{\rho\mu} \qquad\qquad J_{13} = \frac{ff''}{\rho\mu}\frac{\partial\rho\mu}{\partial c} + \beta\left(\frac{\rho_e}{\rho}\right)^2\frac{\partial}{\partial c}\left(\frac{\rho}{\rho_e}\right)$$

$$J_{14} = \frac{\partial\tau'}{\partial T}\frac{\partial T}{\partial f'} + 2f'\beta$$

$$J_{15} = \frac{\partial\tau'}{\partial T}\frac{\partial T}{\partial H} \qquad\qquad J_{17} = -f''$$

$$J_{24} = H, \qquad J_{25} = f', \qquad J_{33} = J\frac{\partial}{\partial c}\left(\frac{Sc}{\rho\mu}\right) - \frac{Scf}{\rho\mu}$$

$$J_{34} = J\frac{\partial}{\partial c}\left(\frac{Sc}{\rho\mu}\right)\frac{\partial T}{\partial f'} \qquad\qquad J_{35} = J\frac{\partial}{\partial T}\left(\frac{Sc}{\rho\mu}\right)\frac{\partial T}{\partial H}$$

$$J_{36} = \frac{Sc}{\rho\mu} \qquad\qquad J_{37} = -c\frac{Sc}{\rho\mu}$$

$$J_{41} = \frac{1}{\rho\mu} \qquad\qquad J_{43} = \frac{f''}{\rho\mu}\frac{\partial\rho\mu}{\partial c}$$

$$J_{44} = \frac{\partial f''}{\partial T}\frac{\partial T}{\partial f'} \qquad\qquad J_{45} = \frac{\partial f''}{\partial T}\frac{\partial T}{\partial H}$$

$$J_{51} = \frac{w_1 f'\,\mathrm{Pr}}{\rho\mu\,\Delta h_o} \qquad\qquad J_{52} = \frac{\mathrm{Pr}}{\rho\mu}$$

$$J_{53} = \frac{\partial}{\partial c}\left(\frac{\mathrm{Pr}}{\rho\mu}\right)w_5 - \frac{\mathrm{Pr}}{\rho\mu\,\Delta h_o}\left\{w_4\frac{\partial\mathrm{Pr}}{\partial c} + w_3\frac{\partial\mathrm{Le}}{\partial c} + w_2\frac{\partial(h_1-h_2)}{\partial c} - fw_6\right\}$$

$$J_{54} = \frac{\partial H'}{\partial T}\frac{\partial T}{\partial f'} - \frac{w_1\tau\,\mathrm{Pr}}{\rho\mu\,\Delta h_o} \qquad\qquad J_{55} = \frac{\partial H'}{\partial T}\frac{\partial T}{\partial H}\frac{\mathrm{Pr}f}{\rho\mu}$$

$$J_{56} = -\frac{\mathrm{Pr}w_6}{\rho\mu\,\Delta h_o} \qquad\qquad J_{57} = \frac{\mathrm{Pr}}{\rho\mu}\left\{H - \frac{cw_6}{\Delta h_o}\right\}$$

$$J_{63} = f' \qquad J_{64} = c \qquad J_{74} = 1$$

where the following quantities have been defined to condense the expressions for the Jacobian elements

$$\Delta h_o = h_{oe} - h_{ow}, \qquad f'' = \tau/\rho\mu$$

$$w_1 = A(\text{Pr}-1)/\text{Pr}, \qquad w_2 = J(\text{Le}-1)/\text{Le}$$

$$w_3 = J(h_1-h_2)/\text{Le}^2, \qquad w_4 = Af'\tau/\text{Pr}^2 \qquad (4.23)$$

$$w_5 = \hat{Q} - fH - [w_1 f'\tau + w_2(h_1-h_2)]\,\Delta h_o$$

$$w_6 = (\text{Le}-1)(h_1-h_2)/\text{Le}$$

The partial derivatives with respect to temperature and stagnation enthalpy ratio are

$$\frac{\partial \tau'}{\partial T} = \frac{ff''}{\rho\mu}\frac{\partial \rho\mu}{\partial T} + \beta\left(\frac{\rho_e}{\rho}\right)^2 \frac{\partial}{\partial T}\left(\frac{\rho}{\rho_e}\right)\frac{\partial T}{\partial f'}$$

$$\frac{\partial H'}{\partial T} = \frac{\partial}{\partial T}\left(\frac{\text{Pr}}{\rho\mu}\right)w_5 - \frac{\text{Pr}}{\Delta h_o \rho\mu}\left\{w_4 \frac{\partial \text{Pr}}{\partial T} + w_3 \frac{\partial \text{Le}}{\partial T} + w_2 \frac{\partial(h_1-h_2)}{\partial T}\right\} \qquad (4.24)$$

$$\frac{\partial f''}{\partial T} = -\frac{f''}{\rho\mu}\frac{\partial \rho\mu}{\partial T}, \qquad \frac{\partial T}{\partial H} = \frac{\partial T}{\partial h}\Delta h_o, \qquad \frac{\partial T}{\partial f'} = -\frac{\partial T}{\partial h}2Af'$$

4.4 APPLICATION OF QUASILINEARIZATION TO ALGEBRAIC INITIAL CONDITIONS

In Chapter 3 we had two initial conditions that were given constants and one, f'', that had to be determined as part of the solution. A homogeneous solution was made with a small initial value for f'' and zeros for the other initial values. This was added to the particular solution to match the boundary condition at the end of the interval.

In the present chapter, we have the two types of initial conditions found in Chapter 3, but we also have the relations among the initial conditions given in Eq. (4.18). These ablation conditions at the wall are algebraic equations relating known and unknown initial boundary conditions and are of the form

$$X_i(0) = F_i[X(0)] \qquad (i = 1,\ldots,n) \qquad (4.25)$$

In this problem, the nonzero elements are

$$F_6 = F_7 = \hat{Q}\,\Delta h_o/[\Delta h_{ab} - (h_1-h_2)(1-c)] \qquad (4.26)$$

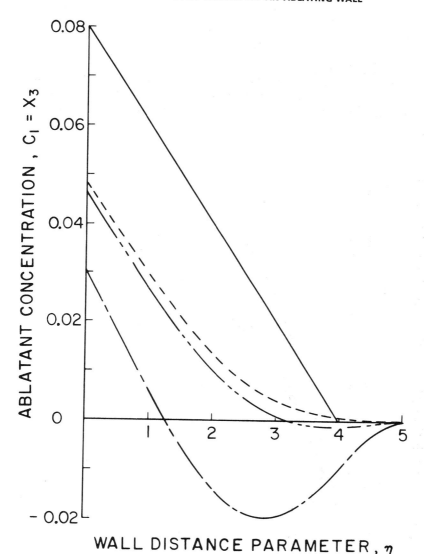

Figure 4.3.   Boundary Layer on an Ablating Wall,
Convergence of Approximations for the
Ablatant Concentration, c.

where

$$\Delta h_{ab} = h_{1w} - h_{1s}$$

is the enthalpy of ablation.

In general the nonzero elements of $\mathbf{F}(\mathbf{X})$ may be nonlinear functions. Here the function is linear in $\hat{Q}$ but nonlinear in c. If Eq. (4.26) is used directly to determine $\hat{J}$ and f in terms of $\hat{Q}$ and c, an inconsistency will result. Let

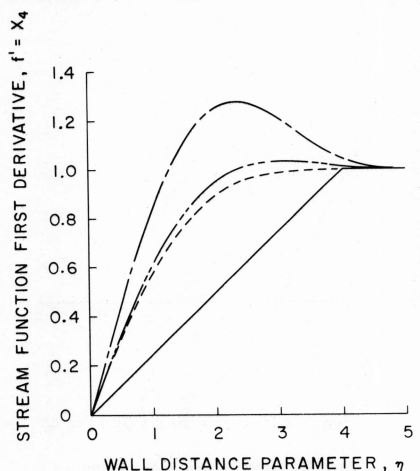

Figure 4.4.   Quasilinearization Solution for the
Boundary Layer on an Ablating Wall, Convergence of
Approximations for the Stream Function First Derivative, f'.

us examine how this occurs. Suppose we take three homogeneous solutions to satisfy the three boundary conditions at the edge of the boundary layer:

$$\mathbf{H}_1 = \{\varepsilon, 0, 0, 0, 0, 0, 0\}$$

$$\mathbf{H}_2 = \left\{0, \varepsilon, 0, 0, 0, \frac{\varepsilon \Delta h_o}{\Delta h_{ab}}, \frac{\varepsilon \Delta h_o}{\Delta h_{ab}}\right\} \tag{4.27}$$

$$\mathbf{H}_3 = \{0, 0, \varepsilon, 0, 0, 0, 0\}$$

where $X_6 = \hat{J}$ and $X_7 = c$ have been calculated from Eq. (4.26). Using

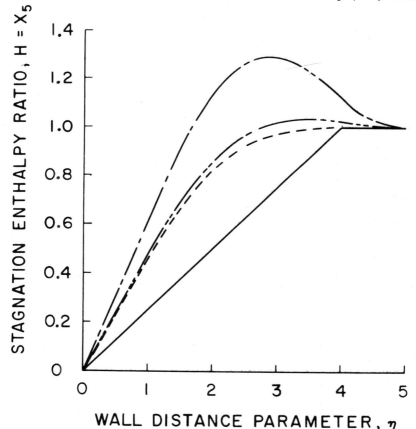

Figure 4.5. Quasilinearization Solution for the Boundary Layer on an ablating Wall, Convergence of Approximations for the Stagnation Enthalpy Ratio, H.

the combination coefficients C_i, obtained as described in Chapter 2, we sum the homogeneous and particular solutions to obtain the new approximation to f and obtain

$$f^{(k+1)} = f_p + C_2 \varepsilon \, \Delta h_o / \Delta h_{ab} \tag{4.28}$$

where f_p is the initial condition for the particular solution. This is not consistent with Eq. (4.26), since a large value of C_3 corresponding to a large change in c will not affect $f^{(k+1)}$.

A logical and consistent approach to the foregoing problem is to apply the same quasilinearization method to the boundary-condition equations as to the differential equations. That is, the boundary-condition equation is expanded about the approximation $\mathbf{X}^{(k)}$ in a Taylor series that is truncated with linear terms. The equations relating the boundary conditions become

$$X_i^{(k+1)}(0) = F_i(\mathbf{X}^{(k)}(0)) + \sum_{j=1}^{N} \frac{\partial F_i(\mathbf{X}^{(k)}(0))}{\partial X_j} [X_j^{(k+1)}(0) - X_j^{(k)}(0)]$$

$$(i = 1, \ldots, n) \tag{4.29}$$

where N is the number of boundary conditions that are not at the initial point and n is the dimension of the state vector. If the succession of approximations converges, the linearized boundary conditions become exact in the limit, in the same manner as the differential equations, because they approach the equations from which they were derived.

Since the initial values of the kth approximation are used to start the particular solution for the $(k+1)$st approximation, we have

$$P_i(0) = F_i(\mathbf{X}^{(k)}(0)) \tag{4.30}$$

For the homogeneous solutions, we have

$$H_{ij}(0) = \frac{\partial F_i(\mathbf{X}^{(k)}(0))}{\partial X_j} \tag{4.31}$$

since we construct $\mathbf{H}_j$, with only one nonzero independent element.

The initial-condition Jacobian J_w has four nonzero terms in the present problem, which are

$$J_{w62} = J_{w72} = \frac{\Delta h_o}{\Delta h_{ab} - (h_1 - h_2)(1-c)}$$

$$J_{w63} = J_{w73} = \frac{-\hat{Q}(J_{w62}-1)(h_1-h_2)}{\Delta h_{ab} - (h_1-h_2)(1-c)} \tag{4.32}$$

The procedure for the initial conditions is, finally, to take three homogeneous solutions as in Eq. (4.27), but with the elements 6 and 7 calculated from Eq. (4.31) and a particular solution taken from the last iteration, with elements 6 and 7 calculated from Eq. (4.30). We shall see below that this method does indeed converge rapidly. In Chapters 5 and 6, we shall see that this method of treating related initial conditions has application to the process of matching a numerical solution to an analytical solution with unknown parameters.

4.5 SAMPLE SOLUTION

A sample solution is presented to illustrate the convergence of the quasilinearization method with three boundary conditions, at the end of the interval, and initial conditions that are related by two algebraic equations. The conditions at the edge of the boundary layer are actually approached asymptotically. However, at least for favorable pressure gradients, the asymptotic boundary conditions are approached so rapidly that the use of analytic solutions at the outer edge of the boundary layer is unnecessary, and the constant conditions are imposed at a finite distance. For the effect of unfavorable pressure gradients, see Libby and Chen [2].

4.5.1 Specification of the Example

The example that we present here is specified by the following choice of parameters:

Molecular weight ratio	$M_{\text{air}}/M_{\text{ablatant}}$	0.2
Specific heat ratio	$C_{p\ \text{ablatant}}/C_{p\ \text{air}}$	2.0
Specific heat ratio	γ	1.3
Temperature ratio	T_w/T_e	0.2
Wedge angle (in units of π radians) $\quad\beta$		0.5
Enthalpy change in ablation	Δh_{ab}	20.00
Lewis number	Le	1.5
Mach number	M	2.0
Prandtl number	Pr	0.8

A fourth-order Runge-Kutta numerical integration method was used with step sizes of 0.2, 0.667, 0.4, and 0.0286 on η, for successive iterations.

Interpolation for values of X_i between the values stored at an interval of 0.2 was carried out with a second difference Stirling formula. The logic to perform these operations is included in the general quasilinearization subroutine, which is described in Chapter 9.

4.5.2 Convergence of Iterations for Boundary-Layer Variables

Convergence of the quasilinearization solution for a two-component boundary layer on an ablating wall is illustrated in Figs. 4.1 through 4.7.

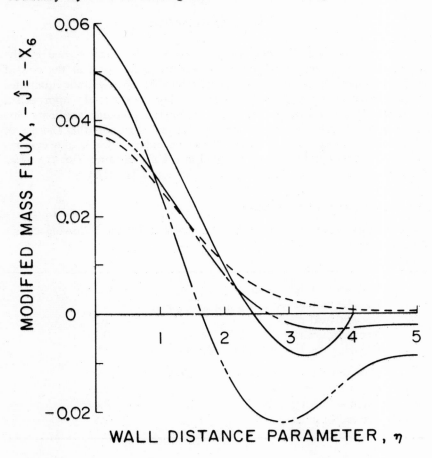

Figure 4.6. Boundary on an Ablating Wall, Convergence of Approximations for the Modified Mass Flux, $-J$.

The initial approximation (solid lines) and the first, second, and fourth (final) iterations are plotted for the 7 variables. The absolute values of the fourth (dashed lines) and third iterations differ by less than 0.01 for any of the variables at any of the 26 storage points. This solution required 23 sec on an IBM 7094 computer.

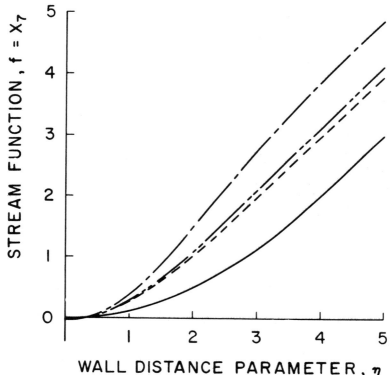

Figure 4.7.   Quasilinearization Solution for the
Boundary Layer on an Ablating Wall, Convergence of
Approximations for the Stream Function, f.

As was the case in Chapter 3 the initial approximation was chosen to be fairly crude to illustrate the convergence of the method. Certainly a prior knowledge of the incompressible boundary-layer solution would provide a better initial approximation than was used. Once one solution has been obtained for one set of parameters, it can profitably be used to initiate solutions for neighboring values of the parameters.

For each of the functions except the modified mass flux $\hat{J}$ (Fig. 4.6), the first iteration overshoots the final solution and the convergence is monotone thereafter. The convergence is not quadratic[†] in the early

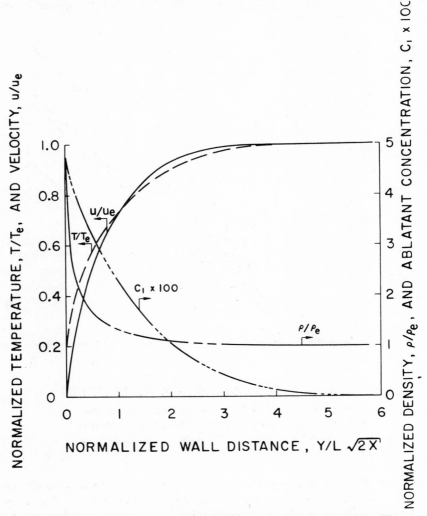

Figure 4.8. "Physical" or Retransformed Boundary Layer Profiles.

[†] In the significant figure-doubling sense.

iterations but is still rapid. When the initial approximation is far from the true solution, the quadratic terms that were dropped in deriving the quasi-linearization algorithm cannot be expected to be small. Hence the monotonicity and doubling of the number of significant figures with each iteration cannot be expected to hold at that stage.

The manner in which the modified mass flux $\hat{J}$ converges is particularly interesting, since its initial condition was derived from c and $\hat{Q}$ through the quasilinearized ablation condition, and the changes are large enough to be seen easily. Although, in the right direction, the initial change is small. However, after $\hat{Q}$ and c, on which the starting value of $\hat{J}$ depends, have made their large initial changes, $\hat{J}$ attains very nearly its final value at the wall on the next iteration. One may infer that variables that have initial conditions derived from nonlinear equations at the beginning of the interval will converge slowly until the other variables from which they are derived have approached their final values.

4.5.3 Retransformation to Physical Profiles

In this problem we have made a transformation of the variables to put the equations into a form that was close to that of a simpler system, for which we possessed a solution. This new form allowed us to make a reasonable initial approximation for some of the variables (although we did not take full advantage of this opportunity in the example). Having obtained convergence to a solution of the transformed problem, we conclude the chapter by retransforming to coordinates that are more simply related to the physical coordinates. The equation (4.7) defining the Dorodnitsyn wall distance parameter η may be inverted to give

$$y/(2\chi)^{1/2} = \int_0^\eta \left(\frac{\rho_e}{\rho}\right) d\eta \qquad (4.33)$$

This quadrature is carried out in an output subroutine at the end of computer program, where ρ/ρ_e is a function of η that may be computed from values of T and c available at each storage point. The "physical" or retransformed boundary-layer profiles of u/u_e, T/T_e, c, and ρ/ρ_e, for the sample solution, are shown plotted on the same abscissa, for comparison, in Fig. 4.8.

4.6 NOMENCLATURE

A parameter in definition of Q

C_i combination coefficient for ith homogeneous solution

c concentration of minor component

D binary diffusion coefficient

$\mathbf{F}$ vector function relating boundary conditions at the wall

f stream function

$\mathbf{g}$ vector function of state vector

$\mathbf{H}$ homogeneous solution

H stagnation-enthalpy ratio

H_{ij} ith component of jth homogeneous solution

h enthalpy

h_o stagnation enthalpy

Δh_{ab} enthalpy increase from solid ablatant to gaseous ablation products

Δh_o difference in stagnation enthalpy of mixture between wall and edge of boundary layer

J mass flux

$\hat{J}$ modified mass flux

J_{ij} element of Jacobian matrix

J_{wij} element of Jacobian matrix for wall boundary-condition equations

k thermal conductivity

L characteristic length

Le Lewis number

M Mach number

m exponent of velocity dependence on x

N number of boundary conditions specified at edge of boundary layer

n number of variables in the state vector

$\mathbf{P}$ particular solution

p pressure

Pr Prandtl number

Q variable in energy equation related to heat flux

$\hat{Q}$ modified heat flux

Re Reynolds number

T temperature of gas

u velocity in x direction (along the body)

v velocity in y direction (normal to surface)

w_j $1 \leqslant j \leqslant 6$, function in definitions of J_{ij}

$\mathbf{X}$ state vector

$X^{(k)}$ kth approximation to state vector

x coordinate in flow direction along body surface

y coordinate perpendicular to body surface

$\hat{y}$ "similar" wall distance parameter $y/(2\chi)^{1/2}$

β equivalent (double) wedge angle of flow in units of π radians

γ ratio of specific heats

η Dorodnitsyn wall distance parameter

μ viscosity

ρ density

τ shear-stress function

χ transformed-length coordinate

ψ stream function

Superscript and subscripts:

e   quantity taken at edge of boundary layer

o   quantity at a reference location or state (except h)

w   quantity taken at wall

$'$ differentiation with respect to η

REFERENCES

1. R. Vaglio-Laurin, "Heat Transfer on Blunt Bodies in General Three-Dimensional Flow," *Heat Transfer Fluid Mechanics Inst.* (1959).
2. P. Libby and Karl K. Chen, Remarks on quasilinearization applied in boundary-layer calculations. *AIAA.* **4** (5), 937 (1966).

Chapter 5

COMPUTATION OF ELECTROSTATIC
PROBE CHARACTERISTICS

5.1 INTRODUCTION

A problem arising in the measurement of properties of plasmas will be exploited in this chapter to illustrate the treatment of a singularity at one end of the interval of integration. The singularity in this case is due to transformation of an infinite interval to a finite interval. An approximate analytical solution valid in the neighborhood of the singularity is used to provide initial conditions for a numerical solution and the analytical solution is quasilinearized at the matching point. Another new feature of this chapter is the determination of parameters as part of the process of solution by utilization of a number of boundary conditions that exceeds the order of the system of differential equations.

5.2 BACKGROUND OF THE PROBLEM

An electrostatic probe in its simplest form is a metal electrode that is placed in a region containing an ionized gas or plasma. The current flowing through the probe to or from the plasma is measured as a function of the applied voltage and properties of the plasma are deduced from this measured relation.

Since large amounts of energy are required to separate positive and negative charges in an ionized gas, the plasma will be neutral except near its boundaries, where electric current may flow to or from the walls of a container or electrodes. Bordering a conducting boundary or electrode is a region of the plasma in which large gradients of electric potential and charge density occur, and which is mathematically analogous to the flowing fluid boundary layer discussed in Chapters 3 and 4.

The distance that the positive and negative charges could be separated in a plasma if all of the thermal energy of the electrons could be converted to potential energy in the electric field is called the Debye length, denoted here by h. This important scale for the plasma sheath is given by

$$h = (kT_e\varepsilon_0/N_0 e^2)^{1/2} \qquad (5.1)$$

where k is the Boltzmann constant, T_e the kinetic temperature of the electrons, ε_0 the permittivity of a vacuum, N_0 the number density of electrons at great distance from a boundary, and e the charge on an electron.

Su and Lam [1] have formulated the general problem of a spherical electrostatic probe in a high-density (collision-dominated), slightly ionized gas and have obtained series solutions for very negative or very small probe potential under the assumption that the probe potential is many Debye lengths. Cohen [2], using the same formulation, has considered the asymptotic limits of large radius-to-Debye length ratio and small ion-to-electron temperature ratio in obtaining other series solutions.

These asymptotic solutions are valuable for obtaining an overall understanding of the problem and for checking numerical solutions. However, a general method of providing numerical solutions for arbitrary values of the parameters of the problem, for comparison with experimental data, is required if the solutions are to contribute to the usefulness of the electrostatic probe as an instrument.

5.3 FORMULATION

The notation used here is derived from that of Su and Lam [1] and Cohen [2]. This formulation assumes that the ionized gas is composed of positive ions, electrons, and neutral atoms. The neutral atoms enter the equations only implicitly through the diffusion coefficients for ions and electrons D_i and D_e. Explicit consideration of three or more species may be required for the description of many plasmas. However, we shall restrict ourselves to the two-species approximation here, since it is adequate to illustrate the application of our method.

In the absence of ionization of neutral atoms and recombination of ions and electrons, a mathematical model of the plasma bordering a probe is given by conservation of ions and electrons:

$$\nabla \cdot [-D_i N_i - (D_i N_i eZ/kT_i)\nabla\phi] = 0$$
$$\nabla \cdot [-D_e N_e + (D_e N_e e/kT_e)\nabla\phi] = 0 \qquad (5.2)$$

and by Poisson's equation:

$$\nabla^2\phi = -(e/\varepsilon_0)(N_i Z - N_e) \qquad (5.3)$$

where the quantities not previously defined are N_i and N_e, the ion and electron number densities; Z, the number of charges on a positive ion; T_i, the ion temperature; and ϕ, the electric potential. Boundary conditions for the spherically symmetric problem considered here are

$$r = r_p: N_i = N_e = 0, \qquad \phi = \phi_p \tag{5.4}$$
$$r \to \infty: \phi \text{ exists and} \to 0, \qquad N_i Z \to N_0, \qquad N_e \to N_0$$

where the subscript p denotes conditions at the probe. A detailed discussion of the boundary conditions is given by Cohen [2]. More recent work by Blue and Ingold [3] indicates that the number density should not be zero at the probe surface, and this effect will be considered in a later section of this chapter.

We specialize the Laplacian to its spherical symmetric form and introduce dimensionless variables and parameters by the relations:

$$\bar{r} = r/r_p, \qquad Y = \phi/\phi_p, \qquad n_e = N_e/N_0, \qquad n_i = ZN_i/N_0$$
$$J_e = I_e/(D_e N_0 4\pi r_p e)$$
$$J_i = I_i/(D_i N_0 4\pi r_p e) \tag{5.5}$$
$$\varepsilon = T_i/T_e Z, \qquad y_p = e\phi_p/kT_e$$
$$\rho_p = r_p/h$$

With this transformation and differentiation with respect to $\bar{r}$ denoted by primes, Eqs. (5.2) and (5.3) become

$$(\bar{r}^2 Y')' = \frac{-\bar{r}^2 \rho_p^2 (n_i - n_e)}{y_p}$$

$$n_i' = \frac{-n_i (y_p/\varepsilon)(\bar{r}^2 Y') + J_i}{\bar{r}^2} \tag{5.6}$$

$$n_e' = \frac{n_e y_p (\bar{r}^2 Y') + J_e}{\bar{r}^2}$$

with the boundary conditions

$$\bar{r} = 1: \quad Y = 1, \qquad n_e = n_i = 0$$
$$\bar{r} \to \infty: \quad Y \to 0, \qquad n_e = n_i \to 1 \tag{5.7}$$

Unlike the fluid dynamic boundary layers in Chapters 3 and 4, the plasma sheath does not approach its asymptotic boundary conditions rapidly beyond some well-defined thickness (e.g., a 99% velocity thickness). In particular, Y approaches zero as $\bar{r}^{-1}$ at large $\bar{r}$. A very large value of $\bar{r}$ is required if the boundary condition is to be imposed at a finite "edge" value to obtain even fair accuracy.

We shall deal with this "edge" problem in three steps: (1) the semi-infinite interval will be transformed to a finite interval; (2) an approximate analytical solution for the neighborhood of the singularity resulting from the transformation will be found; (3) the analytical solution will be evaluated near the edge of its region of validity to provide a boundary condition for the numerical solution.

5.4 TRANSFORMATION TO FINITE INTERVAL

The region between the probe surface at $\bar{r} = 1$ and infinity is mapped into a finite interval by the transformation

$$\zeta = \bar{r}^{-1} \qquad 0 \leqslant \zeta \leqslant 1 \tag{5.8}$$

which converts Eqs. (5.6) to the simpler form

$$Y'' = -\zeta^{-4} \rho_p^2 (n_i - n_e)/y_p$$
$$n_i = -n_i (y_p/\varepsilon) Y' - J_i \tag{5.9}$$
$$n_e' = n_e y_p Y' - J_e$$

and converts the boundary conditions Eqs. (5.7) to

$$\begin{aligned} \zeta = 0: \ &Y = 0, \qquad n_e = n_i = 1 \\ \zeta = 1: \ &Y = 1, \qquad n_e = n_i = 0 \end{aligned} \tag{5.10}$$

5.5 ASYMPTOTIC SOLUTION

In order to avoid numerical integration near the singularity at $\zeta = 0$, an approximate analytical solution is constructed for small ζ. As is often the case, our mathematical approximation is guided by physical insight.

The charged particle flow field is subdivided into a plasma sheath, where the ion and electron densities may differ by a large percentage, and a quasineutral or field penetration region, where the densities are very nearly equal but different from their values at infinity ($\zeta = 0$). These two regions are described more clearly by variables redefined in terms of sums and differences of ion and electron densities and currents:

$$n_t = n_e + n_i, \qquad n_d = n_e - n_i$$
$$J_t = J_e + J_i, \qquad J_d = J_e - J_i \tag{5.11}$$

With primes now denoting differentiation with respect to ζ, Eqs. (5.9) become

$$Y'' = C n_d \zeta^{-4}$$
$$n'_t = Y'(A n_t + B n_d) - J_d \tag{5.12}$$
$$n'_d = Y'(B n_t + A n_d) - J_t$$

where

$$A = y_p(1 - \varepsilon^{-1})/2, \qquad B = y_p(1 + \varepsilon^{-1})/2, \qquad C = \rho_p^2/y_p$$

and the boundary conditions (5.10) become

$$\zeta = 0: \ Y = 0, \qquad n_d = 0, \qquad n_t = 2$$
$$\zeta = 1: \ Y = 1, \qquad n_d = 0, \qquad n_t = 0 \tag{5.13}$$

The form that the solution must approach asymptotically as ζ approaches zero is found by noticing that, in this region, the charge density is "quasineutral," i.e., That $n_d \approx 0$. Setting $n_d = 0$ and substituting the third of Eqs. (5.12) into the second yields

$$n_t = 2 + \zeta(A/B)J_t - J_d = 2 + \zeta \Delta J \tag{5.14a}$$

Substituting this back into the third of Eqs. (5.12) and integrating the normalized potential, Y is found to be

$$Y = (BJ_t/\Delta J)\ln(n_t/2) \tag{5.14b}$$

Boundary conditions for a numerical solution over the range $\zeta_0 \leqslant \zeta \leqslant 1$ can be obtained by evaluating Eqs. (5.14a) and (5.14b) at ζ_0. We chose ζ_0 to make n_d suitably small at that point. If n_d grows rapidly in the numerical portion of the solution just beyond ζ_0, then ζ_0 must be decreased.

The solution of Eqs. (5.14) and (5.15) has been obtained without the use of the first of Eqs. (5.12) (Poisson's equation), so that it is a solution to a system of reduced order. As a result the solution contains only two parameters, J_t and J_d, which cannot be adjusted to satisfy three boundary conditions at the probe surface ($\zeta = 1$). A third parameter is obtained by integrating the first of Eqs. (5.12) from $\zeta = 0$ to ζ, to give

$$Y' = S, \qquad Y = S\zeta \qquad (5.15)$$

where S is a constant. This solution corresponds to the field outside of a spherical charge distribution with no current flowing. The converged solution can be expected to have the quasineutral character of the first asymptotic solution in Eqs. (5.14) and (5.15). However, the intermediate approximations may have some of the charge penetration characteristics of Eq. (5.15), since the approximate charge distribution will result in incomplete shielding of the electrode.

Since Eqs. (5.12) are nonlinear, we might expect that the two asymptotic solutions above could not be superimposed. However, since the solution of Eq. (5.15) is of the nature of a small perturbation, the addition of the solutions is consistent with the order of our approximation. Even if this were not so, the addition could be carried out after the linearization that will be effected in Section 5.8. In that section, it will be shown that S need not actually be treated as a parameter.

The two parameters J_t and J_d are initially unknown and must be determined as part of the solution. Let us consider next a simple method for determining unknown parameters as part of a boundary-value problem.

5.6 TREATMENT OF PARAMETERS

A simple method for determining unknown parameters is to treat them as variables. Although this procedure is inefficient when actually applied, it has the advantage of requiring no special procedure to handle parameters. For the parameters of the present problem, we write

$$J_t'(\zeta) = 0, \qquad J_t(\zeta_0) = J_t$$
$$J_d'(\zeta) = 0, \qquad J_d(\zeta_0) = J_d \qquad (5.16)$$
$$S'(\zeta) = 0, \qquad S(\zeta_0) = S$$

When Eqs. (5.16) are added to Eqs. (5.12), a seventh-order set of ordinary differential equations results, with seven prescribed boundary conditions.[†]

[†] Four of the "boundary conditions" are equations relating variables at a boundary.

Since the order matches the number of prescribed boundary conditions, the methods of Chapter 4 are applicable.

5.7 QUASILINEARIZATION

The problem is now in the proper form for the application of quasilinearization. Expressing Eqs. (5.12) and (5.16) as a system of six[†] first-order equations, and employing the usual state vector notations

$$\mathbf{X}' = \mathbf{g}(\mathbf{X}, \zeta)$$

we have, for the state vector,

$$\mathbf{X} = \{Y', J_t, J_d, Y, n_t, n_d\} \tag{5.17}$$

and, for the vector function $\mathbf{g}$,

$$\mathbf{g} = \{\zeta^{-4} C n_d, 0, 0, Y', Y'(An_t + Bn_d) - J_d, Y'(Bn_t + An_d) - J_t\} \tag{5.18}$$

In this problem the Jacobian matrix is particularly simple:

$$\begin{bmatrix}
0 & 0 & 0 & 0 & 0 & C\zeta^{-4} \\
0 & 0 & 0 & 0 & 0 & 0 \\
0 & 0 & 0 & 0 & 0 & 0 \\
1 & 0 & 0 & 0 & 0 & 0 \\
An_t + Bn_d & 0 & -1 & 0 & AY' & BY' \\
Bn_t + An_d & -1 & 0 & 0 & BY' & AY'
\end{bmatrix} \tag{5.19}$$

Recalling the vectors defining the boundary conditions, which were introduced in Eq. (3.11), the vectors $\mathbf{B}$, $\mathbf{L}$, and $\mathbf{M}$ are written:

$$\mathbf{B} = \{1, 0, 0\}, \qquad \mathbf{L} = \{1, 1, 1\} \tag{5.20}$$

$$\mathbf{M} = \{4, 5, 6\}$$

The initial conditions have not been included because they are satisfied,

---

† We shall show in Section 5.8 that S can be treated as a boundary condition only.

in the limit, by the way the numerical solution is started. This starting
method is examined in the next section.

5.8 ASYMPTOTIC SOLUTION AS BOUNDARY CONDITIONS

The asymptotic solution presented in Section 5.5 constitutes a set of
nonlinear algebraic equations

$$X_i^{(k+1)}(\zeta_0) = F_i[X^{(k+1)}(\zeta_0), \zeta_0] \qquad i = 1, \ldots, 6 \qquad (5.21)$$

relating the boundary conditions at the matching point ζ_0 to the three
parameters J_t, J_d, and S. This is the same type of boundary condition that
holds at the wall in the ablating boundary-layer problem in Chapter 4.

To insure that the improved solutions formed by summing the homogeneous and the particular solutions satisfy the boundary conditions to the
same order as the differential equations, the boundary conditions are again
linearized in the same manner as in Section 4.4 [see Eq. (4.29)].

Integration could be carried out from $\zeta = 1$ to $\zeta = \zeta_0$; however, four
conditions are specified by the asymptotic solution at the matching point ζ_0
and only three are specified at the probe. Thus, integration from ζ_0 to the
probe surface is advantageous because one less homogeneous solution is
required. Of equal importance is the possibility of using the same computer
program structure as that in Chapter 4, if Eq. (5.21) are used as initial
conditions. By superimposing Eqs. (5.14) and (5.15), the function **F** in Eq. (5.21)
is found to be

$$\mathbf{F} = \left\{ \begin{array}{c} J_t/B(2+\zeta_0\Delta J)+S \\[1ex] J_t \\[1ex] J_d \\[1ex] (J_t/B\Delta J)\ln(1+\zeta_0\Delta J/2)+S\zeta_0 \\[1ex] 2+\zeta_0\Delta J \\[1ex] 0 \end{array} \right\} \qquad (5.22)$$

From the arguments in Section 5.5, we may now expect that the terms
containing S will be small compared with the quasineutral terms and will
vanish when the solution converges. However, in order to approach the
final solution, we require three homogeneous solutions to match the three
boundary conditions at $\zeta = 1$ (the probe surface). These three homogeneous

solutions must be based on variations of three parameters in the asymptotic solution. We notice that $S = Y'(\zeta_0)$, so that S need not be treated as a separate parameter but can be treated as a variable with an unknown initial condition. (See f'' in Chapter 3). Thus S is replaced by Y' and Eq. (5.15) produces two entries in the Y' column of the Jacobian matrix:

$$
\left[\frac{\partial F_i}{\partial X_j}\right] =
\begin{bmatrix}
1 & \dfrac{1 - Y_q'\zeta_0 A}{Bn_t} & Y_q'\zeta_0 & 0 & 0 \\[2ex]
0 & 1 & 0 & 0 & 0 \\[1ex]
0 & 0 & 1 & 0 & 0 \\[1ex]
\zeta_0 & \dfrac{Y_q' A\zeta_0 - Y_q(J_d/J_t)B}{B\Delta J} & \dfrac{Y_q - \zeta_0 Y_q'}{\Delta J} & 0 & 0 \\[2ex]
0 & A\zeta_0/B & -\zeta_0 & 0 & 0 \\[1ex]
0 & 0 & 0 & 0 & 0
\end{bmatrix}
\qquad (5.23)
$$

where Y_q and Y_q' are from Eqs. (5.14) and the $\partial F_1/\partial X_1 = 1$ is derived from Eq. (5.15).

In the numerical computations for the results shown later in this chapter, the terms containing S were neglected in Eq. (5.22) as was the $\partial F_4/\partial X_3$ element in Eq. (5.23). Convergence may have been slowed by these simplifications, but it was still very rapid, as will be shown in Section 5.10.

5.9 ORTHOGONALIZATION OF HOMOGENEOUS SOLUTIONS

The homogeneous solutions to the quasilinearized equations of this problem tend to become dependent as ζ increases, and a special procedure must be instituted to insure that the linear equations, which must be solved for the combination coefficients, are independent. Consider a matrix representing a vector space in which the (column) vectors $\mathbf{H}_j$ correspond to homogeneous solutions of the quasilinearized equation (5.12) and the rows correspond to the variables X_i. By selecting the rows connected with the variables that must satisfy boundary conditions at the final point (i.e., X_4, X_5, X_6 at $\zeta = 1$ in this example), a square matrix $\hat{H}$ is obtained, representing a subspace of the solution space. This matrix, evaluated at $\zeta = 1$, appears in the equation

$$
B_i - P_i(1) = \sum_{j=1}^{3} C_j H_{ij}(1), \qquad (i = 4, 5, 6) \qquad (5.24)
$$

which must be solved for the combination coefficients C_j of the homogeneous solutions $\mathbf{H}_j$ in the new approximation. Even though the $\mathbf{H}_j$ are initially orthogonal, they may become parallel as ζ increases toward 1, corresponding to the determinant $|\hat{H}_{ij}|$ approaching zero. In this event the C_j cannot be found.

The tendency of the matrix $\hat{H}$ to become increasingly ill-conditioned as the integration proceeds is associated in many cases with an exponentially growing solution that dominates all of the homogeneous solution vectors.

Exponential behavior of homogeneous solutions to the quasilinearized differential equations is to be expected from the form of equations. Consider, e.g., the derivative of a variable X_i whose right-hand side depends on X_i among other variables in the form

$$X_i'^{(k+1)} = \frac{\partial g_i}{\partial X_i}(\mathbf{X}^{(k)}(\zeta))X_i^{(k+1)} + \sum_{\substack{j=1 \\ j \neq i}}^{n} \frac{\partial g_i}{\partial X_j}(\mathbf{X}^{(k)}(\zeta))X_j^{(k+1)} \tag{5.25}$$

Integrating formally, we obtain

$$X_i^{(k+1)} = \exp\left(\int_0^\zeta \frac{g_i}{X_i}\,d\zeta\right)\exp\left(\int_0^\zeta \sum_{\substack{j=i \\ j \neq 1}}^{n} \frac{\partial g_i}{\partial X_j} X_j^{(k+1)}/X_i^{(k+1)}\,d\zeta\right) \tag{5.26}$$

If $(\partial g_i/\partial X_i)(X^{(k)}(\zeta))$ is a slowly varying function of ζ, and if, for the other elements of that row of the Jacobian matrix, $\partial g_i/\partial X_j \ll \partial g_i/\partial X_i$, then

$$X_i^{(k+1)}(\zeta) \approx \exp^{(\partial g_i/\partial X_i)\,\zeta} \tag{5.27}$$

Of course, the relations are not this simple, even in the present problem, but we are led to expect rapidly growing particular and homogeneous solutions.

Figure 5.1 shows schematically the effect of the rapidly growing component of the solution. The initial homogeneous solution vectors are shown along the coordinate axes. The homogeneous solution vectors at a larger value of the independent variable are represented by the longer, nearly parallel set of vectors. At this point the Gramm-Schmidt orthogonalization procedure [4] is applied to produce a new orthogonal set of vectors. This new set does not lie along the axes.

As a rough estimate of the condition of the matrix $\hat{H}$, the ratio of the volume of the parallelepiped, formed by the homogeneous solution vectors, to the volume defined by an orthogonal set of vectors of equal length is taken. This quantity is calculated at each point where the solution is stored for the next iteration and compared with a numerical criterion. In the present problem, the criterion becomes: Orthogonalize if

$$|\hat{H}| \, / \prod_{j=1}^{3} |\hat{H}_j| < 0.001 \tag{5.28}$$

For application of a more elegant condition number, see Newman and Todd [5].

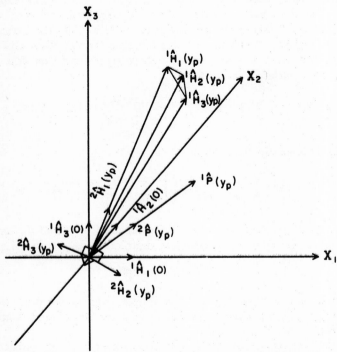

Figure 5.1. Orthogonalization of Homogeneous Solutions.

Since the orthogonalization procedure may have to be applied several times in the range of integration, the solution is broken into segments that are denoted by a presuperscript, e.g., $^{k}H(\zeta)$, the homogeneous solution matrix in the kth segment. The transformation ^{k}T, which orthogonalizes the subspace vectors at the end of the kth segment, is applied to the whole matrix ^{k}H to obtain the transformed matrix

$$^{k+1}H = \,^{k}H^{k}T \tag{5.29}$$

containing consistent state vectors $^{k+1}H_j$, so that the integration may be continued.

Since the magnitude of the vectors before transformation can be quite large, it is desirable to use an orthonormal transformation. In order to prevent loss of significance in the calculation of the combination coefficients, the component $\hat{P}$ of the particular solution in the subspace also is transformed to a unit vector by

$$^{k+1}P = {}^kP - {}^{k+1}H\,{}^kC_T \tag{5.30}$$

where

$$^kC_T = {}^k\hat{P}\,{}^{k+1}\hat{H}(1 - |{}^k\hat{P}|^{-1}) \tag{5.31}$$

5.10 DISCUSSION OF COMPUTER SOLUTIONS

Convergence of the quasilinearization method is illustrated in Figs. 5.2a and 5.2b. The former is for a small probe and small relative potential ($\rho_p = 1$, $y_p = 1$), and the latter is for a large probe and large relative potential ($\rho_p = 100$, $y_p = 8$). In both cases the rate of convergence is extremely rapid, with nearly all of the change in the final shape accomplished

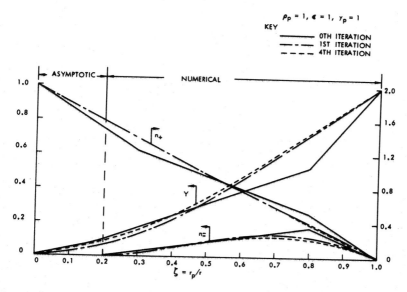

Figure 5.2a. Convergence of Quasilinearization Process.

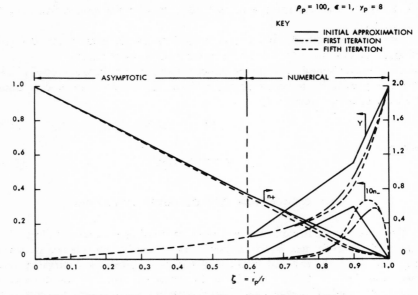

Figure 5.2b.   Convergence of Quasilinearization Process.

in the first iteration. Note that the values of the asymptotic solution at the matching point change as the numerical solution converges.

Solutions are shown parametrically in Figs. 5.3, 5.4a, 5.4b, and 5.5. In all cases the solutions may be divided into a sheath region and a quasineutral, or ambipolar, diffusion region. In the sheath region the normalized probe potential Y drops rapidly; there is a maximum in the normalized difference in charge carrier densities n_d; and the slope of the sum of the charge carrier densities n_t is reduced. In the quasineutral region the asymptotic solution is applied; n_d approaches zero; n_t is linear in ζ; and the potential falls slowly.

In Figure 5.3 the normalized potential Y changes very little with y_p, although n_d increases and n_t decreases in the sheath region. At $y_p = 5$ the nearly linear shape of n_t at lower potentials is skewed to produce a region of very small slope in the center of the sheath, corresponding to nearly constant charge carrier density.

The effect of increasing dimensionless probe radius ρ_p on the solution is seen, from Figs. 5.4a and 5.4b, to be opposite to increasing potential y_p on n_t and n_d. The normalized potential distribution Y becomes steeper, and the maximum of n_d is shifted toward the probe surface, corresponding to a thinner sheath relative to probe radius.

The effect of the third parameter ε, which is the ratio of ion and electron temperatures divided by the number of charges per positive ion, is shown in Fig. 5.5. The sum of the charge carrier densities has only a small change with increasing ε, but the effect on Y and n_d is similar to decreasing ρ_p.

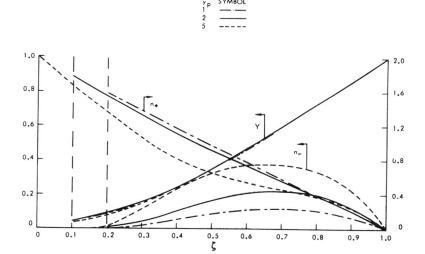

Figure 5.3. Dependence of Profile on Dimensionless Probe Potential, y_p.

Figure 5.6 shows the dimensionless currents J_e and J_i as a function of the dimensionless probe potential y_p with the dimensionless probe radius ρ_p as a parameter for $\varepsilon = 1$. Each solution of Eqs. (5.12) yields one pair of characteristic values J_t and J_d, resulting in one pair of points in Fig. 5.6. Only solutions with y_p positive were required to construct Fig. 5.6, since, for $\varepsilon = 1$, the system of equations is invariant under the transformation

$$y_p \rightarrow -y_p, \quad n_e \rightarrow n_i, \quad n_i \rightarrow n_e$$
$$J_e \rightarrow -J_i, \quad J_i \rightarrow -J_e \tag{5.32}$$

All of the curves pass through 1 on the J axis, as can be shown from the equations. At the axis the curvature of J_e is zero, going negative slowly

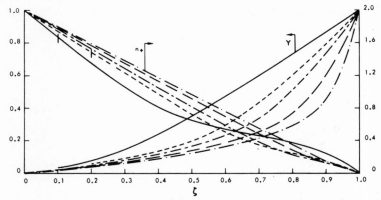

Figure 5.4a.   Dependence of Profile Shape on Dimensionless Probe Radius, ρ_p.

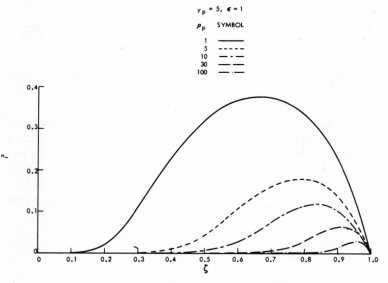

Figure 5.4b.   Dependence of Profile Shape on Dimensionless Probe Radius, ρ_p.

with y_p; so the curves finally roll off toward a constant slope, proportional to $(y_p/\rho_p)^{1/2}$ [1]. To the left of the J axis, the curvature becomes positive immediately, so that the intercept is approximately an inflection point. This feature is important in the data reduction method discussed in the original paper by Radbill [6]. At large negative y_p, the J_e curves approach the y_p axis asymptotically.

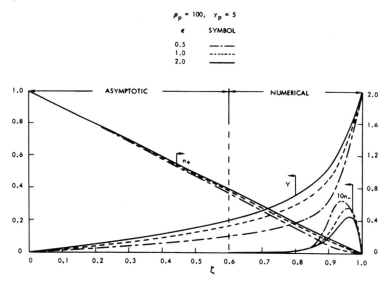

Figure 5.5.   Dependence of Profile Shape on $\varepsilon = T_i/T_e Z$.

The asymptotic solution of Cohen [2] is plotted, for comparison, for three values of ρ_p. The quasilinearization and asymptotic solutions for J_e agree well at $\rho_p = 100$, where they are directly comparable. The agreement for J_i is less than for J_e. The curves for $\rho_p = 50$ and $\rho_p = 30$ appear to be consistent in shape and location.

In order to compare the analytical, numerical results with experimental data, total current must be known as a function of probe potential. The diffusion terms in Eq. (5.2) insure that, in the collision-dominated plasma considered, both positive and negative particles are collected by any electrode at moderate potential. Thus, the separate measurement of ion and electron currents does not appear feasible.

The ratio of the electron and ion diffusion coefficient D_e/D_i, which does not appear in the dimensionless equations, must be known or assumed to construct curves of dimensionless total current from the type of information given in Fig. 5.6. In air at atmospheric pressure, this ratio may be about 5000 [7], but if negative ions are formed that tie up the electrons,

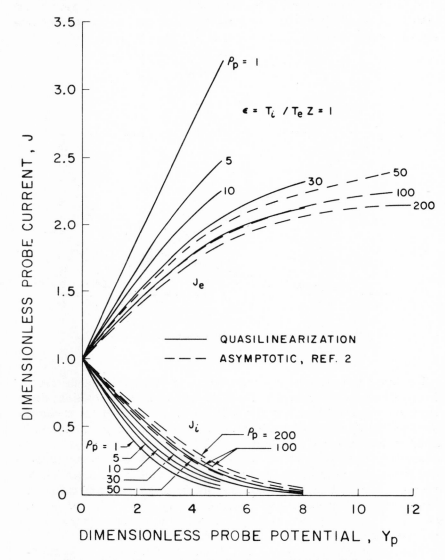

Figure 5.6.   Electrostatic Probe Current-Voltage Relation.

the ratio may be much smaller. In the stagnation region of a hypersonic vehicle or wind-tunnel model, where the concentration of contaminants or ablatants is uncertain, an *a priori* assumption of D_e/D_i appears unjustified.

Figure 5.7 shows total dimensionless current J curves constructed with D_e/D_i equal to 10, 100, and 1000. Even for 100, the J curve is barely distinguishable from the J_e curve. For D_e/D_i equal to 10, the portion of the

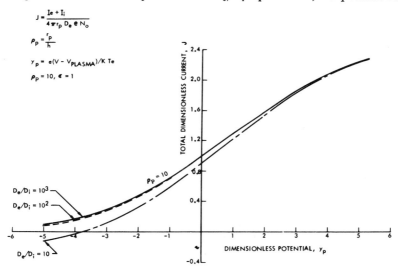

Figure 5.7. Effect of Diffusion Coefficient Ratio D_e/D_i on Total Dimensionless Current, J.

curve to the left of the J axis is shifted noticeably, although the portion to the right and the intercept on the J axis are changed very little. Thus, the electron current J_e can be substituted for the total current J in the first quadrant without reference to the diffusion coefficient ratio because the error is, at most, D_i/D_e. The intercept on the axis of dimensionless potential y_{pi} (floating potential) may be deduced, from the analysis of Su and Lam [1], to be $y_{pi} \approx \ln (D_i/D_e \varepsilon)$ for $D_i/D_e \ll 1$.

5.11 COUPLED BOUNDARY CONDITIONS AT BOTH ENDS OF THE INTERVAL

The situation may arise in which the boundary conditions on a set of ordinary differential equations are related by equations at both ends of the integration interval. Boundary conditions of this type are found in the present problem if the more realistic boundary conditions derived by Blue and Ingold [3] are substituted for the simpler conditions formulated by Su and Lam [1]. These new boundary conditions state that the particle fluxes and particle densities are linearly related at the probe surface. In terms of the dimensionless currents and densities, we have

$$J_e = a_e n_e, \qquad J_i = a_i n_i, \qquad \zeta = 1 \tag{5.33}$$

where a_e and a_i were 11.1 and -213, respectively, in the case studied by Blue and Ingold [8]. Converting to the variables defined in Eq. (5.11), there results at $\zeta = 1$:

$$n_t = \{[(a_e)^{-1}+(a_i)^{-1}]J_t+[(a_e)^{-1}-(a_i)^{-1}]J_d\}/2$$

$$n_d = \{[(a_e)^{-1}-(a_i)^{-1}]J_t+[(a_e)^{-1}+(a_i)^{-1}]J_d\}/2 \tag{5.34}$$

$$Y = 1$$

Although these are linear relations between the variables at the end of the interval, for the sake of generality, we shall consider the treatment of nonlinear relations. Suppose that the boundary conditions on the components of the state vector at the end of the interval ζ_f are related by the vector equation

$$\mathbf{X}(\zeta_f) = \mathbf{h}(\mathbf{X}(\zeta_f)) \tag{5.35}$$

where some of the $h_i(\mathbf{X})$ may be nonlinear and some may be zero. In order to be consistent with the linearized differential equations and the linearized initial conditions, Eq. (5.35) must also be linearized in the same manner to give

$$X_i^{(k+1)} = h_i(\mathbf{X}^{(k)}) + \sum_{j=1}^{n} \frac{\partial h_i}{\partial X_j}(\mathbf{X}^{(k)})[X_j^{(k+1)} - X_j^{(k)}] \tag{5.36}$$

where all the dependent variables are evaluated at ζ_f, and n is the dimension of the state vector.

Now from Chapter 2, the new approximation $X_i^{(k+1)}$ is given in terms of the particular and homogeneous solutions by

$$X_i^{(k+1)} = P_i + \sum_{j=1}^{M} C_j H_{ij} \qquad (i = 1, \ldots, n) \tag{5.37}$$

where M is the number of homogeneous solutions, and the C_j are determined by satisfying the boundary conditions.

Upon combining Eqs. (5.36) and (5.37) with some rearrangement, we obtain

$$\sum_{m=1}^{M} C_m \sum_{j=1}^{n} \left[\delta_j^i - \frac{\partial h_i}{\partial X_j} (\mathbf{X}^{(k)}) \right] H_{jm} = -P_i + h_i(\mathbf{X}^{(k)})$$

$$- \sum_{j=1}^{n} \frac{\partial h_i}{\partial X_j} (\mathbf{X}^{(k)}) [X_j^{(k)} - P_j] \qquad (i = 1, 2, \ldots, M) \qquad (5.38)$$

where M is the number of boundary conditions that are not initial conditions and n is the dimension of the state vector. This set of M linear equations in the M unknowns C_m are solved at the end of each iteration (i.e., $\zeta = 1$).

In the notation of Eq. (5.35), Eq. (5.34) becomes

$$\begin{aligned} h_5 &= AX_2(1) + BX_3(1) \\ h_6 &= BX_2(1) + AX_3(1) \end{aligned} \qquad (5.39)$$

and the nonzero elements of the Jacobian matrix $[\partial h_i/\partial X_j]$ become

$$\partial h_5/\partial X_2 = A, \qquad \partial h_5/\partial X_3 = B, \qquad \partial h_6/\partial X_2 = B$$

$$\partial h_6/\partial X_3 = A$$

where

$$A = [(a_e)^{-1} + (a_i)^{-1}]/2$$

$$B = [(a_e)^{-1} + (a_i)^{-1}]/2$$

The capability of handling dependent boundary conditions at both ends of the integration interval is not included in the computer program presented in the appendix. An older and simpler version of that program was modified, by addition of a subroutine, to calculate the h_i and $\partial h_i/\partial X_j$ and by addition of logic, to implement Eq. (5.38) prior to solution of the linear equations.

5.11.1 Convergence with Coupled Boundary Conditions at Both Ends of the Interval

The quasilinearization method applied to a problem with coupled boundary conditions at both ends was found to converge rapidly. Several values of y_p were run at $\varepsilon = 1$ and $\rho = 127$, in order to allow comparison with computations made by Blue and Ingold [3] by a different method. Convergence of the case $y_p = 5$ is shown in Fig. 5.8 where the first, second, and fifth iterations are shown. The analytical portion of the solution is not shown in order to emphasize that the "boundary conditions" change as the process of solution proceeds. The changes in the boundary conditions at $\zeta = \zeta_0 = 0.6$ are relatively large, whereas the change in n_t at $\zeta = 1$ are small due to a good initial approximation. At this fairly low value of $y_p(y_p = 5)$, the profiles of Y and n_t resemble their counterparts in Fig. 5.2 through 5.5, except for the nonzero value of n_t at $\zeta = 1$. However, n_d does not show the relatively large maximum near $\zeta = 1$ found in the previous solutions.

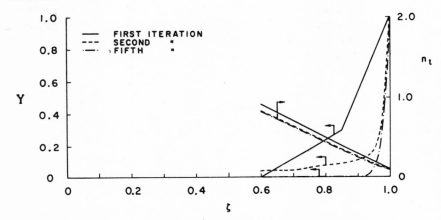

Figure 5.8. Convergence with Coupled Boundary Conditions at Both Ends of Interval — Low Probe Potential.

A solution for a large probe potential $y_p = 15$ is shown in Fig. 5.9, where both the analytical and numerical portions of the solution are shown. Both the sum n_t and difference n_d of the charge carrier densities are affected in an important manner by the coupled boundary conditions at the wall. The total charge carrier density n_t actually shows a minimum near the probe surface. The charge carrier difference n_d has an inflection point, where it is nearly constant over a small region and then rises to its maximum value at the boundary.

In spite of the coupled boundary conditions at both ends of the interval, the solution to the electrostatic probe problem converged rapidly. The additional complication of the coupled boundary condition at the probe surface ($\zeta = 1$) has proved worth investigating because it has produced a qualitatively different solution. An important difference in eigenvalues also occurs, for which the reader is referred to Blue and Ingold [3]. We observe the importance of proper boundary conditions in problems involving nonlinear differential equations. What appeared to be a minor change in boundary conditions in Eq. (5.33) has led to a qualitatively different solution.

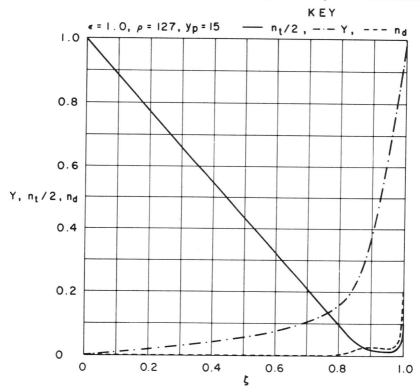

Figure 5.9. Convergence with Coupled Boundary Conditions at Both Ends of Interval — High Probe Potential.

REFERENCES

1. C. H. Su and L. H. Lam, Continuum theory of spherical electrostatic probes. *Phys. Fluids* 6, 1479–1491 (October, 1963).
2. I. M. Cohen, Asymptotic theory of spherical electrostatic probes in a slightly ionized, collision-dominated gas. *Phys. Fluids* 6, 1492–1499 (October, 1963).
3. E. Blue and J. H. Ingold, "The Theory of Langmuir Probe Characteristics for a Collision Dominated Partially Ionized Gas." Gen. Elec. Co. Tech. Rept. AFAL-TR-67-1 (February, 1967).

4. F. B. Hildebrand, "Methods of Applied Mathematics," pp. 34–35. Prentice-Hall, Englewood Cliffs, New Jersey, 1952.
5. M. Newman and J. Todd, The evaluation of matrix inversion programs. *J. Soc. Ind. Appl. Math.* **6** (4), 466–476 (1958).
6. J. R. Radbill, Computation of electrostatic probe characteristics by quasilinearization, *AIAA J.* **4** (7), 1195–1200, p. 201 (1966).
7. L. B. Loeb, "Basic Processes of Gaseous Electronics," p. 201. Univ. of California Press, Berkeley, California, 1961.
8. E. Blue, Private communication, 1966.

Chapter 6

PREDICTION OF THE STABILITY OF LAMINAR BOUNDARY LAYERS

6.1 INTRODUCTION

A characteristic value problem of large dimension is examined in the present chapter. The Orr-Sommerfeld equation that governs the problem is equivalent to a linear fourth-order complex system of equations. When the two eigenvalues and two matching parameters are considered as variables, the equations become nonlinear and the dimension of the state vector is increased to 12. An analytical asymptotic solution valid for large values of the independent variable is employed to provide boundary conditions for the numerical solution at the outer edge of the boundary layer in a manner analogous to the use of the quasineutral solution in Chapter 5.

The occurrence of a large parameter in the equation leads to a rapidly growing parasitic solution, again analogous to Chapter 5, that dominates the homogeneous solutions and requires their orthogonalization, as well as scaling of the particular solution. In order to secure convergence of the quasilinearization algorithm, the concept of restriction of initial variation in parameter space is introduced. The eigenvalues are restrained for the first few iterations until the eigenfunctions have approached a self-consistent form.

Extensive plots are presented to aid in visualizing the solution surface in parameter space and in visualizing the physical significance of the self-excited disturbance in the laminar boundary layer on a flat plate. Although the primary objective of this chapter is to demonstrate the application of quasilinearization to the characteristic value problem exemplified by boundary-layer stability, an important secondary objective is to convey a clear understanding of the nature of the Tollmien-Schlichting instability to the reader not intimately involved in the field of fluid mechanics. Because the primary emphasis is mathematical rather than historical, space allows mention of only a few significant or representative contributions to the extensive literature. Those wishing a more complete bibliography are referred to recent review articles by Mack [1] and Reid [2].

6.2 BACKGROUND

The laminar boundary layer, which was studied in Chapter 3, is subject to self-excited instabilities at length Reynolds numbers above a certain critical value. These instabilities are waves that propagate at a fraction of the free-stream velocity and may grow or damp exponentially with time and/or distance. The early stages of the dispersive propagation of these Tollmien-Schlichting waves are accurately described by the solutions of the Orr-Sommerfeld equation, which we shall study in later sections. As the disturbances grow, they become nonlinear and may themselves become subject to higher order instabilities, as has been suggested by Eckhaus [3]. In some manner these nonlinearities and additional instabilities are believed to lead to a complete breakdown of the laminar flow that results in turbulence.

Since the experiments of Reynolds [4] toward the end of the last century, a large number of authors have investigated the transition from laminar to turbulent flow. Tollmien [5] was the first to develop an effective theoretical method to study small disturbances in laminar flows that were presumed to lead to turbulence. Tollmien's asymptotic theory, expanded by other authors, was the classical method of solution of the hydrodynamic stability problem until the advent of large digital computers. Among the extensions of Tollmien's theory, the work of Schlichting ([6] and [7]) on the Blasius boundary layer (on a flat plate) is particularly complete, including eigenvalues and eigenfunctions.

Final experimental proof of the physical existence of self-sustaining oscillations induced within the boundary layer was carried out by Schubauer and Skramstad [8] at the U.S. Bureau of Standards during World War II and published in 1947. These experiments established complete confidence in the basic oscillatory theory.

Lin [9] has placed the asymptotic theory on a firm mathematical basis, clarifying some of the more difficult details, such as the behavior near singular points in the expansion and the choices of proper branches of multiple valued functions. By means of this work, Lin and more recently Shen [10], have improved the accuracy of stability computations made from the Tollmien theory.

As modern digital computers become available, a variety of numerical methods become feasible. Three of these methods are briefly described, and comparison is made with the application of quasilinearization in later sections of this chapter.

The method of Kurtz and Crandall [11] converts the Orr-Sommerfeld ordinary differential equation to a finite-difference equation that results in a large set of linear algebraic equations, one for each increment in the wall distance. The eigenvalues are found by searching for zeros of the determinant in an indirect manner. The determinant is evaluated on a grid of values of the parameters α and Re, whereupon the zeros are found by interpolation. In the quasilinearization approach, however, convergence to eigenvalues is an automatic process.

Step-by-step integration methods are used by both Kaplan [12] and Nachtsheim [13]. The former starts at the edge of the boundary layer with an analytical solution of an asymptotic form of the differential equation, and integrates toward the wall. The latter starts at the wall and integrates outward. Kaplan satisfies initially only the two conditions at the wall derived from the tangential velocity and uses a series of solutions with interpolation on the eigenvalue parameters to satisfy finally the normal velocity condition and to obtain the eigenvalues. Nachtsheim utilizes the Newton-Raphson method to find the zeros of the deviation of his solution from the boundary condition at the outer edge of the boundary layer in terms of the eigenvalue parameters. In contrast the present method, which is equivalent to a generalized Newton-Raphson method, satisfies all of the boundary conditions in every iteration while satisfying the differential equation only in the limit.

In the present chapter, which is derived from Radbill and van Driest [14], the quasilinearization algorithm is applied to compute accurately the eigenfunctions and associated eigenvalues for the stability of the Blasius incompressible laminar boundary layer. Details of the flow fields for particular sets of parameters determining the disturbance and the mean flow are shown as well.

6.3 THE ORR-SOMMERFELD EQUATION

The growth of small disturbances in a two-dimensional incompressible laminar boundary layer is governed by the Orr-Sommerfeld equation, which may be derived from the Navier-Stokes equations and continuity. The derivation may be found in standard texts such as those of Schlichting [15] and Pai [16]. Principle assumptions in the derivation are that the boundary layer may be approximated by a parallel flow and that quadratic terms in the perturbation can be neglected. In addition the perturbation is assumed to have a stream function ψ' of the form

$$\psi' = \phi(y) \exp\left[i(\alpha x - \beta t)\right] \tag{6.1}$$

where ψ', ϕ, and β are complex. The wave number α is defined as $2\pi/\lambda$

wherein λ is the wave length, and β is the wave frequency. The Orr-Sommerfeld equation is accordingly

$$i\alpha \text{Re}((U-c)(\phi''-\alpha^2\phi)-U'') = \phi^{iv}-2\alpha^2\phi''+\alpha^4 \qquad (6.2)$$

where $c = \beta/\alpha$ is the complex propagation velocity, and differentiation by y is denoted by primes. The boundary conditions are

$$y = 0: \phi = \phi' = 0; \qquad y \to \infty: \phi \text{ and } \phi' \text{ are bounded}$$

6.4 SOLUTION FOR LARGE Y

Although one boundary condition is imposed at infinity, an analytical solution valid for large y may be found, so that the numerical integration may be carried out over a finite interval. The analytical solution is used to provide boundary conditions for the numerical solution in the same manner as in Section 5.8.

The boundary conditions on the basic parallel flow, which approximates the boundary layer and which has a free-stream velocity which is constant with x, are

$$y \to \infty: U \to 1, \qquad U'' \to 0$$

Thus, for $y \geqslant y_0$, where y_0 is sufficiently large, Eq. (6.2) reduces to

$$\phi^{iv}-(2\alpha^2 + i\alpha\text{Re}(1-c))\phi'' + (\alpha^4 + i\alpha^3\text{Re}(1-c))\phi = 0 \qquad (6.3)$$

Upon application of the standard method for solution of linear ordinary differential equations with constant coefficients to Eq. (6.3) and retention of the two solutions that are bounded at infinity, there results

$$\phi(y) = A_1 \exp(-\alpha y) + A_2 \exp(fy) \qquad (6.4)$$

where A_1 and A_2 are complex constants to be determined from the numerical solution, and

$$f = -(\alpha^2 + i\alpha\text{Re}(1-c))^{1/2} \qquad (6.5)$$

Since the magnitude and phase of the solution of a homogeneous linear differential equation are arbitrary, we set $A_1 = 1$. Furthermore, in order to avoid calculation of exponentials with large arguments, the second term in Eq. (6.4) is scaled through multiplication and division by $\exp(f_0 y)$, where f_0 is calculated from an initial approximation to the solution. The solution for large y becomes

$$\phi(y) = \exp(-\alpha y) + A \exp(fy - f_0 y_0) = \phi_1 + A\phi_2 \qquad (6.6)$$

where A, from which the subscript 2 has been dropped, must be determined from the numerical solution.

Both Kaplan [12] and Nachtsheim [13] use this solution as a boundary condition at large y (Kaplan as an initial and Nachtsheim as a final condition). Kurtz and Crandall [11], however, appear to keep only the first term, probably because they do not deal with the derivatives explicitly where the second term becomes important.

6.5 QUASILINEARIZATION

Although the Orr-Sommerfeld equation is linear in the y-dependent part of the stream function $\phi(y)$ and its derivatives, which are the variables of the problem, the equation is nonlinear in the parameters and eigenvalues, which we shall vary in the process of solution. The state vector for this problem is

$$\mathbf{X} = \{\phi''', \phi'', \phi', \phi, \begin{bmatrix} \alpha\mathrm{Re}, & \alpha^2 \\ c_r, & c_i \\ c_r, & \alpha^2 \end{bmatrix}, A\} \tag{6.7}$$

where three alternate choices of parameters have been shown and primes denote differentiation with respect to y. The Orr-Sommerfeld equation may be written as a vector equation with 4 complex first-order components in the form

$$\mathbf{X}' = \mathbf{g}(\mathbf{X}) \tag{6.8}$$

where

$$\mathbf{g}(\mathbf{X}) = \{(i\alpha\,\mathrm{Re}(U-c)+2\alpha^2)\phi'' - (i\alpha\,\mathrm{Re}(\alpha^2(U-c)+U'')+\alpha^4)\phi, \ \phi''', \phi'', \phi'\}$$

Depending on the choice of independent parameters in the problem, the parameter portion of the state vector may have two complex components or one complex component and two real components that are not the real and imaginary components of a complex number. In either case the state vector has the equivalent of 12 real components.

In order to reduce the nonlinearity of the dependence on the parameters in Eq. (6.8) and to simplify the Jacobian matrix $[\partial g_i/\partial X_j]$, the parameters Re and α are replaced by $\alpha\mathrm{Re}$ and α^2. These, together with c_r and c_i, constitute four real parameters in Eq. (6.8), two of which must be specified and two (eigenvalues) must be determined in the process of solution. Two choices of eigenvalues were used in the computations, namely, $(\alpha\mathrm{Re}, \alpha^2)$ and (c_r, α^2).

The pair of eigenvalues (c_r, c_i) was not used, since curves of constant c_i were desired. However, this last pair constitutes a complex number, and some saving in computation can be realized by its use, as will be noted below. The search for higher modes of disturbance in the related problem of Chapter 7 is also aided by using (c_r, c_i) as eigenvalues.

The Jacobian matrix $[\partial g_i/\partial X_j]$ for the first choice of eigenvalues is

$$
\begin{bmatrix}
0 & i\alpha\mathrm{Re}(U-c) & 0 & -i\alpha\mathrm{R}_e(\alpha^2(U-c) & i((\phi''-\alpha^2\phi)(U-c)\,2\phi''-\phi(i\alpha\mathrm{Re}(U-c) \\
 & +2\alpha^2 & & +U'')+(\alpha^2)^2 & -U''\phi) & +2\alpha^2) \\
1 & 0 & 0 & 0 & 0 & 0 \\
0 & 1 & 0 & 0 & 0 & 0 \\
0 & 0 & 1 & 0 & 0 & 0
\end{bmatrix}
$$

For the second choice of eigenvalues, the fifth column is replaced by

$$\{-i\alpha\mathrm{Re}(\phi''-\alpha^2\phi),\ 0,\ 0,\ 0\}$$

We recall that by requiring the particular solution **P** to satisfy the initial conditions, only as many homogeneous solutions $\mathbf{H}_j$ as there are boundary conditions at the second point are required. In this case we shall integrate from the edge of the boundary layer at y_0 toward the wall at $y = 0$ where there are four boundary conditions, so that four homogeneous solutions are required. If the solution process is carried out in terms of complex variables, with c as the single complex eigenvalue, only two complex homogeneous solutions are required to satisfy the two complex boundary conditions at the wall. However, if eigenvalues are selected that are not real and imaginary parts of a complex number, three complex homogeneous solutions are needed. In the computations described in this chapter, the state vector was treated as composed of real components, and four homogeneous solutions were used. Although this treatment consumed additional computation time, it avoided making extensive changes in well-checked-out computer programs that were available.

Together with the homogeneous boundary conditions following Eq. (6.2), Eq. (6.8) constitutes an eigenvalue problem. If two of the four parameters, α^2, $\alpha\mathrm{Re}$, c_r, and c_i are specified, the remaining two become eigenvalues to be determined. Each solution may be represented as a point on an amplification surface composed of curves of constant c_i in a space whose coordinates are α^2, $\alpha\mathrm{Re}$, and c_r. An equivalent surface with coordinates α, Re, and c_r is described and shown below. Other combinations of parameters can be used in an analogous manner.

We note that although these eigenvalues are treated as search "variables" in the quasilinearization process and are obtained at the end of each iteration by summation of homogeneous and particular solutions, they are not functions of y. Thus, the eigenvalues are not actually integrated on y, as was suggested in Chapter 5, but are handled separately.

In the computation of the Jacobian matrix $[\partial g_i/\partial X_j]$ in this section and in Chapter 6, we require the contribution to a Jacobian matrix from the differentiation of the real and imaginary parts of one complex variable X_m by the real and imaginary parts of another complex variable X_n. The motivation of this procedure is found in the necessity of coupling a subroutine, which uses complex arithmetic, to a general purpose program written in terms of real variables. A typical element of the 4-element matrix block is

$$\frac{\partial X_{mi}}{\partial X_{ni}} = \mathrm{Im}\left\{\frac{\partial X_m}{\partial X_{ni}}\frac{\partial X_n}{\partial X_{ni}}\right\} = -\mathrm{R}\left\{\frac{\partial X_m}{\partial X_n}\right\}$$

and the whole block becomes:

	X_{nr}	X_{ni}
X_{mr}	$\mathrm{R}\left\{\dfrac{\partial X_m}{\partial X_n}\right\}$	$-\mathrm{Im}\left\{\dfrac{\partial X_m}{\partial X_n}\right\}$
X_{mi}	$\mathrm{Im}\left\{\dfrac{\partial X_m}{\partial X_n}\right\}$	$-\mathrm{R}\left\{\dfrac{\partial X_m}{\partial X_n}\right\}$

where subscripts r and i have been used for real and imaginary parts. Of course, if the quasilinearization algorithm is programmed in terms of complex variables, this subterfuge is unnecessary and the 4-element block in the Jacobian matrix is replaced by one complex element.

6.6 QUASILINEARIZATION OF SOLUTION FOR LARGE Y

As was previously stated in Section 6.4, where the asymptotic solution was derived, this solution will be used to provide initial conditions for the numerical solution that is integrated toward the wall from the outer edge of the boundary layer. The solution for large y, Eq. (6.6), is nonlinear in the parameters A, α^2, and αRe (or c_r), which are varied in the process of solution. In order to be consistent with the differential equations, the boundary conditions given by Eq. (6.6) and its first three derivatives written in the form:

$$F_n = (-\alpha)^{4-n}\phi_1 + Af^{4-n}\phi_2 \qquad (n = 1, 2, 3, 4) \qquad (6.9)$$

must be linearized as was done in Chapters 4 and 5 to yield

$$X_n^{(k+1)}(y_0) = F_n(\mathbf{X}^{(k)}, y_0) + \sum_{j=1}^{4} \frac{\partial F_n(\mathbf{X}^{(k)}, y_0)}{\partial X_j} (X_j^{(k+1)} - X_j^{(k)}) \qquad (6.10)$$

The nonzero elements of the Jacobian matrix $[\partial F_n/\partial X_j]$ of the initial conditions are

$$\frac{\partial F_{4-n}}{\partial X_5} = \begin{cases} h(n)\,i(1-c) & X_5 = \alpha\text{Re} \\ -h(n)\,i\alpha\text{Re} & X_5 = c_r \end{cases}$$

$$\frac{\partial F_{4-n}}{\partial X_6} = (n - \alpha y_0)(-\alpha)^{n-2}\phi_1 + h(n) \qquad (6.11)$$

$$\frac{\partial F_{4-n}}{\partial X_7} = f^n \phi_2$$

where $h(n) = (n + fy_0)A\phi_2 f^{n-2}/2$, $n = 1, 2, 3, 4$ and $i = (-1)^{1/2}$.

Since the integration is carried out in the negative y direction, the signs of all the elements of $\mathbf{g}$ and $[\partial g_i/\partial X_j]$ must be changed in the actual computation. However, the signs of the elements of F and $[\partial F_n/\partial X_j]$ are not changed, since they are not explicitly derivatives with respect to y.

6.7 ORTHOGONALIZATION OF HOMOGENEOUS SOLUTIONS

The orthogonalization technique described in Chapter 5 was an essential element of the solution of Eq. (6.2). That equation possesses a very rapidly growing solution that dominates the homogeneous solutions after only a few integration steps. In the first few iterations, 25 orthogonalizations were required, which decreased to about 5 when the process converged. The growth of the unwanted solution is associated with the large parameter αRe in Eq. (6.2) and becomes more rapid as that parameter is increased.

6.8 METHOD OF NUMERICAL ANALYSIS
AND COMPUTER SOLUTION

During the quasilinearization iteration process, intermediate and final solutions were stored at an interval of 0.2 from $y = 0$ to $y - y_0 = 8$. The functions $U(y)$ and $U''(y)$, which represent the undisturbed Blasius velocity profile, are also stored at the same interval. Quadratic interpolation was carried out using the Newton forward difference formula in the first interval, the Newton backward difference formula in the last interval, and the Bessel interpolation in the interior intervals [17]. Integration was started using the fourth-order Gill modifications of the Runge-Kutta (see [18]) method and continued using the Adams-Moulton integration [19], retaining second differences.

In general a smaller step size is required for the integration formula than for the interpolation formula if they are of the same order. The integration is performed on the particular and homogeneous solutions, $\mathbf{P}$ and $\mathbf{H}_j$, which many have large higher derivatives, particularly in the early stages of convergence. On the other hand, the interpolation is performed on tabulated values of the kth approximation to the solutions $\mathbf{X}^{(k)}$ which are usually much smoother than $\mathbf{P}$ and $\mathbf{H}_j$. The unwanted rapidly growing solutions that necessitated the orthogonalization in Section 6.7 tend to be cancelled out by the summing of $\mathbf{P}$ and the $\mathbf{H}_j$, leaving $\mathbf{X}^{(k)}$ with much smaller higher derivatives.

The eigenfunction ϕ was not very sensitive to the Reynolds number; therefore, the change between neighboring solutions was small and the convergence was essentially accomplished in one step. However, if the initial guess for the eigenvalue is poor, the eigenfunction might fail to converge. The region of convergence appeared to be increased when the eigenvalues were fixed at the initial guess for the first iteration while the eigenfunction approached its final form. The eigenvalues were then freed on succeeding iterations.

The quasilinearization process was usually allowed to run until the profiles of ϕ and its derivatives at each storage point, as well as the eigenvalues, agreed to 0.0001 between successive iterations, and convergence usually occurred in five cycles. The running time was about 1 min. per case.

As has been stated in previous sections, two versions of the program, corresponding to two choices of the pair of eigenvalues, have been used. A particular choice of eigenvalues appeared to work best in regions of parameter space where the search was carried out most nearly perpendicular to the surface formed by curves of constant c_i. Near the maximum value of c_i, $\alpha\delta^*\mathrm{Re}^*$ and c_i were fixed, whereupon c_i and $(\alpha\delta^*)^2$ were allowed to vary. Near the minimum value of $\alpha\delta*\text{Re}_\delta*$, c_r and c_i were fixed, whereupon c_i and $(\alpha\delta*)^2$ were allowed to vary. In regions such as those just below the minimum value of $\alpha\delta*\text{Re}_\delta*$ on the lower branch, both programs worked equally well. Other pairs of eigenvalues could have been used, such as c_r and c_i, with $(\alpha\delta*)^2$ and $\alpha\delta*\text{Re}_\delta*$ fixed.

6.9　COMPARISON OF QUASILINEARIZATION WITH OTHER NUMERICAL METHODS

The quasilinearization method is between the method of Kurtz and Crandall [11], on the one hand, and the method used by Kaplan [12], as well as Nachtsheim [13], on the other hand, in the amount of intermediate storage required. The integration accuracy of the difference equation method used by Kurtz and Crandall is proportional to h^4 where h is the step size. Hence, the results must be stored at a large number of points. In the program that implements the present method described in Chapter 9, the previous iteration results are not stored at every iteration step but rather obtained by interpolation at intermediate points. The difference method has the disadvantage of having to solve a large number of linear equations, whereas the present method, although iterative, has to solve only a few linear equations. Both these methods have the advantage of carrying a check on the convergence of the solution at points in the interior of the interval, whereas the methods of Kaplan and Nachtsheim check convergence only at the end of the interval. The methods that store interior values always satisfy the boundary conditions, but satisfy the equation only in the limit, in contrast to the last two "shoot and correct" methods that always satisfy the equation and try to satisfy the second boundary condition in the limit.

Both the Nachtsheim paper and the present paper use a Newton-Raphson technique, although the former applies the technique only to the values at the end of the interval, while the quasilinearization method applies the technique at every integration step.

6.10　EIGENFUNCTIONS AND EIGENVALUES

The boundary-layer stability computer program calculates and prints out the real and imaginary parts of the eigenfunction, their first three derivatives, and the corresponding eigenvalues. Eq. (6.1) shows that the eigenfunction $\phi(y)$ is the y-dependent factor of the perturbation stream function and that its first derivative is the y-dependent factor of the perturbation velocity parallel to the surface.

Figure 6.1 shows two examples of the eigenfunction ϕ corresponding to the upper branch of the neutral curve ($c_i = 0$) at $Re_\delta{}^* = 2080$ and the lower branch of the neutral curve at $Re_\delta{}^* = 902$, where $Re_\delta{}^*$ is the Reynolds number in terms of boundary layer displacement thickness δ^*. Associated

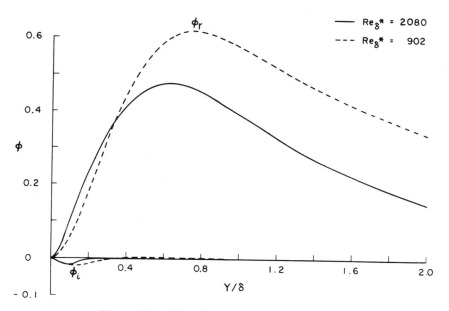

Figure 6.1. Eigenfunctions, $\phi = \phi_r + i\phi_i$, Corresponding to Data of Schubauer and Skramstad for the Neutral Stability Curve at $Re_{\delta*} = 902$ (Lower Branch) and $Re_{\delta*} = 2080$ (Upper Branch).

plots of $d\phi/dy$ are shown in Fig. 6.2. These particular solutions were chosen because they correspond to the experimental velocity fluctuation data of Schubauer and Skramstad [8]. They are tabulated in Tables I and II. The eigenfunctions agree qualitatively with those of Kurtz and Crandall [11], although they are of different Reynolds number and wave number.

The above examples are typical of the upper and lower branches of the neutral curve, or of regions of small damping or amplification close to the neutral curve. It is seen in Fig. 6.1 that the imaginary or out-of-phase

Table I. Eigenfunction Corresponding to Data of Schubauer and Skramstad[a] on Upper Branch of Neutral Curve

$$Re_{\delta*} = 2078.15 \qquad \alpha\delta^* = 0.31228 \qquad c_r = 0.33052 \qquad c_i = 0$$

y	y/δ	ϕ_r	ϕ_i	$d\phi_r/dy$	$d\phi_i/dy$	$d^2\phi_r/dy^2$	$d^2\phi_i/dy^2$
0.0	0.000	0.00000	0.00000	0.00000	0.00000	1.42270	−1.35360
0.2	0.039	0.02454	−0.01142	0.21671	−0.06210	0.61924	0.25332
0.4	0.077	0.07546	−0.01774	0.27394	−0.00172	0.05788	0.25486
0.6	0.116	0.13019	−0.01429	0.26954	0.03003	−0.06262	0.07002
0.8	0.155	0.18273	−0.00775	0.25530	0.03174	−0.07799	−0.03877
1.0	0.194	0.23208	−0.00251	0.23734	0.01960	−0.10435	−0.07147
1.2	0.232	0.27726	0.00005	0.21356	0.00674	−0.13051	−0.05116
1.4	0.271	0.31730	0.00062	0.18685	0.00010	−0.13229	−0.01621
1.6	0.310	0.35211	0.00048	0.16176	−0.00089	−0.11776	0.00273
1.8	0.349	0.38220	0.00038	0.13948	−0.00001	−0.10655	0.00418
2.0	0.387	0.40799	0.00044	0.11855	0.00048	−0.10387	0.00082
2.2	0.426	0.42962	0.00054	0.09774	0.00045	−0.10420	−0.00075
2.4	0.465	0.44709	0.00061	0.07696	0.00028	−0.10326	−0.00085
2.6	0.504	0.46044	0.00065	0.05658	0.00012	−0.10020	−0.00073
2.8	0.542	0.46978	0.00067	0.03701	−0.00001	−0.09514	−0.00064
3.0	0.581	0.47532	0.00065	0.01865	−0.00013	−0.08824	−0.00053
3.2	0.620	0.47734	0.00062	0.00182	−0.00022	−0.07977	−0.00040
3.4	0.659	0.47617	0.00056	−0.01318	−0.00029	−0.07013	−0.00026
3.6	0.697	0.47220	0.00050	−0.02618	−0.00033	−0.05981	−0.00014
3.8	0.736	0.46584	0.00044	−0.03710	−0.00034	−0.04933	−0.00001
4.0	0.775	0.45750	0.00037	−0.04593	−0.00033	−0.03914	0.00007
4.2	0.814	0.44760	0.00030	−0.05280	−0.00031	−0.02964	0.00013
4.4	0.852	0.43650	0.00024	−0.05785	−0.00028	−0.02111	0.00018
4.6	0.891	0.42456	0.00019	−0.06132	−0.00024	−0.01373	0.00020
4.8	0.930	0.41207	0.00015	−0.06343	−0.00020	−0.00758	0.00020
5.0	0.969	0.39926	0.00011	−0.06443	−0.00016	−0.00263	0.00018
5.2	1.007	0.38635	0.00008	−0.06455	−0.00013	0.00122	0.00017
5.4	1.046	0.37349	0.00006	−0.06400	−0.00010	0.00411	0.00014
5.6	1.085	0.36078	0.00004	−0.06296	−0.00007	0.00618	0.00011
5.8	1.124	0.34833	0.00003	−0.06158	−0.00005	0.00758	0.00009
6.0	1.162	0.33617	0.00002	−0.05996	−0.00004	0.00848	0.00007
6.2	1.201	0.32435	0.00001	−0.05821	−0.00003	0.00900	0.00005
6.4	1.240	0.31289	0.00001	−0.05638	−0.00002	0.00923	0.00003
6.6	1.278	0.30180	0.00000	−0.05453	−0.00001	0.00927	0.00002
6.8	1.317	0.29108	0.00000	−0.05268	−0.00001	0.00918	0.00002
7.0	1.356	0.28072	0.00000	−0.05087	0.00000	0.00900	0.00001
7.2	1.395	0.27073	0.00000	−0.04909	0.00000	0.00877	0.00000
7.4	1.433	0.26108	0.00000	−0.04736	0.00000	0.00852	0.00001
7.6	1.472	0.25178	0.00000	−0.04568	0.00000	0.00825	0.00000
7.8	1.511	0.24281	0.00000	−0.04406	0.00000	0.00797	0.00000
8.0	1.550	0.23415	0.00000	−0.04249	0.00000	0.00770	0.00000

Table I. Eigenfunction Corresponding to Data of Schubauer and Skramstad[a] on Upper Branch of Neutral Curve (Cont.)

$Re_{\delta*} = 2078.15$		$\alpha\delta^* = 0.31228$		$c_r = 0.33052$		$c_i = 0$	
y	y/δ	ϕ_r	ϕ_i	$d\phi_r/dy$	$d\phi_i/dy$	$d^2\phi_r/dy^2$	$d^2\phi_i/dy^2$
8.2	1.588	0.22581	0.00000	−0.04098	0.00000	0.00744	0.00000
8.4	1.627	0.21776	0.00000	−0.03952	0.00000	0.00717	0.00000
8.6	1.666	0.21000	0.00000	−0.03811	0.00000	0.00692	0.00000
8.8	1.705	0.20251	0.00000	−0.03675	0.00000	0.00667	0.00000
9.0	1.743	0.19529	0.00000	−0.03544	0.00000	0.00643	0.00000

[a] G. B. Schubauer and H. K. Skramstad, *Natl. Bur. Standards Res. Paper 1172;* also *J. Aeron. Sci.* **14,** 69 (1947).

Table II. Eigenfunction Corresponding to Data of Schubauer and Skramstad[a] on Lower Branch of Neutral Curve

$Re_{\delta*} = 897.93$		$\alpha\delta^* = 0.18182$		$c_r = 0.33156$		$c_i = 0$	
y	y/δ	ϕ_r	$\phi_{i\alpha}$	$d\phi_r/dy$	$d\phi_i/dy$	$d^2\phi_r/dy^2$	$d^2\phi_i/dy^2$
0.0	0.000	0.00000	0.00000	0.00000	0.00000	0.53597	−0.45121
0.2	0.039	0.01031	−0.00576	0.09959	−0.04411	0.43592	−0.04483
0.4	0.077	0.03794	−0.01416	0.17190	−0.03501	0.29115	0.10489
0.6	0.116	0.07737	−0.01880	0.21890	−0.01081	0.18632	0.12296
0.8	0.155	0.12437	−0.01870	0.24884	0.01056	0.11715	0.08574
1.0	0.194	0.17611	−0.01521	0.26670	0.02261	0.06244	0.03486
1.2	0.232	0.23035	−0.01030	0.27393	0.02514	0.00995	−0.00684
1.4	0.271	0.28499	−0.00559	0.27082	0.02118	−0.04023	−0.02931
1.6	0.310	0.33805	−0.00199	0.25841	0.01468	−0.08196	−0.03307
1.8	0.349	0.38788	0.00032	0.23893	0.00871	−0.11045	−0.02547
2.0	0.387	0.43334	0.00163	0.21511	0.00467	−0.12576	−0.01506
2.2	0.426	0.47379	0.00232	0.18926	0.00250	−0.13161	−0.00731
2.4	0.465	0.50900	0.00270	0.16281	0.00147	−0.13225	−0.00356
2.6	0.504	0.53893	0.00294	0.13652	0.00089	−0.13033	−0.00259
2.8	0.542	0.56365	0.00306	0.11080	0.00037	−0.12661	−0.00268
3.0	0.581	0.58331	0.00308	0.08601	−0.00018	−0.12090	−0.00272
3.2	0.620	0.59815	0.00299	0.06259	−0.00069	−0.11298	−0.00240
3.4	0.659	0.60847	0.00281	0.04096	−0.00112	−0.10300	−0.00181
3.6	0.697	0.61468	0.00256	0.02150	−0.00141	−0.09145	−0.00113
3.8	0.736	0.61723	0.00225	0.00444	−0.00157	−0.07901	−0.00049
4.0	0.775	0.61662	0.00193	−0.01010	−0.00161	−0.06638	0.00005

Table II. Eigenfunction Corresponding to Data of Schubauer and Skramstad[a] on Lower Branch of Neutral Curve (Cont.)

$Re_{\delta*} = 897.93$		$\alpha\delta^* = 0.18182$		$c_r = 0.33156$		$c_i = 0$	
y	y/δ	ϕ_r	ϕ_i	$d\phi_r/dy$	$d\phi_i/dy$	$d^2\phi_r/dy^2$	$d^2\phi_i/dy^2$
4.2	0.814	0.61336	0.00162	−0.02214	−0.00156	−0.05415	0.00046
4.4	0.852	0.60792	0.00132	−0.03181	−0.00144	−0.04279	0.00074
4.6	0.891	0.60077	0.00104	−0.03933	−0.00127	−0.03264	0.00090
4.8	0.930	0.59231	0.00081	−0.04496	−0.00108	−0.02390	0.00096
5.0	0.969	0.58289	0.00061	−0.04899	−0.00089	−0.01660	0.00094
5.2	1.007	0.57281	0.00045	−0.05170	−0.00071	−0.01070	0.00086
5.4	1.046	0.56229	0.00032	−0.05335	−0.00055	−0.00608	0.00075
5.6	1.085	0.55152	0.00023	−0.05420	−0.00041	−0.00258	0.00063
5.8	1.124	0.54064	0.00016	−0.05445	−0.00030	0.00000	0.00050
6.0	1.162	0.52977	0.00010	−0.05425	−0.00021	0.00184	0.00039
6.2	1.201	0.51896	0.00007	−0.05375	−0.00015	0.00311	0.00028
6.4	1.240	0.50828	0.00004	−0.05304	−0.00010	0.00395	0.00021
6.6	1.278	0.49776	0.00003	−0.05219	−0.00006	0.00448	0.00013
6.8	1.317	0.48741	0.00002	−0.05126	−0.00004	0.00477	0.00009
7.0	1.356	0.47726	0.00001	−0.05029	−0.00003	0.00491	0.00006
7.2	1.395	0.46730	0.00001	−0.04930	−0.00002	0.00497	0.00004
7.4	1.433	0.45754	0.00000	−0.04831	−0.00001	0.00497	0.00003
7.6	1.472	0.44797	0.00000	−0.04732	−0.00001	0.00493	0.00002
7.8	1.511	0.43861	0.00000	−0.04634	0.00000	0.00486	0.00001
8.0	1.550	0.42944	0.00000	−0.04537	0.00000	0.00478	0.00001
8.2	1.588	0.42046	0.00000	−0.04443	0.00000	0.00469	0.00001
8.4	1.627	0.41167	0.00000	−0.04350	0.00000	0.00460	0.00000
8.6	1.666	0.40306	0.00000	−0.04259	0.00000	0.00450	0.00000
8.8	1.705	0.39463	0.00000	−0.04170	0.00000	0.00441	0.00000
9.0	1.743	0.38638	0.00000	−0.04083	0.00000	0.00431	0.00000

[a] G. B. Schubauer and H. K. Skramstad, *Natl. Bur. Standards Res. Paper 1172*; also *J. Aeron. Sci.* **14**, 69 (1947).

part of ϕ is important only near the wall, whereas the real part of ϕ reaches a maximum considerably farther out in the flow and then decreases approximately as $\exp(-\alpha y)$. Also, note in Fig. 6.2 that the oscillation in the boundary layer is closer to the wall for the upper branch than the lower branch. Except for this relative position, the x-velocity $(d\phi_r/dy)$ curves for both branches are generally of the same shape.

Reference to Figs. 6.1 and 6.2 shows that the wave is not confined to the boundary layer, i.e., $y < \delta$, but extends many thicknesses into the inviscid region with a slow exponential decrease. Because of this extension outside the boundary layer, there could be coupling with an external disturbance in this region if one were imposed.

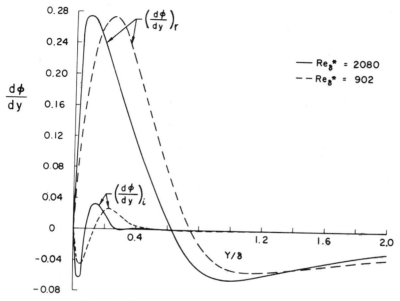

Figure 6.2.   Eigenfunction First Derivative,
$d\phi/dy = (d\phi/dy)_r + i(d\phi/dy)_i$, **Corresponding to Data of
Schubauer and Skramstad for the Neutral Stability Curve at $Re_{\delta*} = 902$
(Lower Branch) and $Re_{\delta*} = 2080$ (Upper Branch).**

Furthermore, from Fig. 6.2, the maximum magnitude of $d\phi_i/dy$ is small compared to that of $d\phi_r/dy$. Although the imaginary part is small, it will be shown later to be important in the mechanism of the disturbance.

The parameters in the combination used in the quasilinearization solution are shown in Figs 6.3, 6.4, and 6.5. These figures constitute three projections of an amplification surface in a space whose coordinates are $\alpha\delta* Re_{\delta*}$, $(\alpha\delta*)^2$, and c_r. As was noted earlier, the use of these combinations simplified the Jacobian matrix.

Reference to these figures shows clearly that the choice of the eigenvalues on which the iterative search is carried out is important. The chosen parameters must define a plane that intersects the desired curve on the surface. For example, starting from an initial estimate point below the minimum Reynolds number for the neutral curve ($c_i = 0$) a search on c_r and $(\alpha\delta^*)^2$ will miss the neutral curve. In practice the eigenvalues will oscillate and no convergence will be obtained. On the other hand, a search on $\alpha\delta^*\mathrm{Re}^*$ and c_r will locate the neutral curve, provided that $(\alpha\delta^*)^2$ is less than the maximum value for the neutral curve.

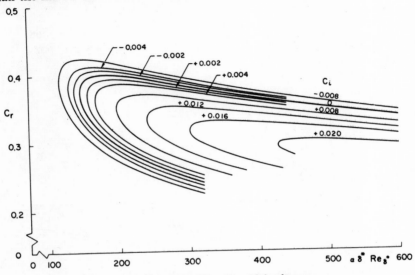

Figure 6.3. Propagation Velocity, c_r, as a Function of Wave Number Reynolds Number Product, $\alpha\,\delta^*\,\mathrm{Re}_{\delta^*}$, for Various Amplification Factors, c_i.

In cases where two different choices of search plane intersect the parameter surface, the plane that cuts the surface most nearly at right angles appears to give the most rapid convergence.

The parameter of "amplification" surface is shown in terms of the more conventional variables Re_δ^*, $\alpha\delta^*$, and c_r in oblique projection in Fig. 6.6. In addition to curves of constant amplification number c_i[†], which lie in the surface and may be either positive or negative, traces of planes of constant Re_δ^*, $\alpha\delta^*$, and $\beta_r v/U^2$ are shown. The surface is illustrated as a solid, cut by a cylinder through the $c_i = 0.008$ curve and parallel to the $\beta_r v/U^2$ axis. This aids in visualizing the surface. The eigenvalues on which this figure is based are tabulated in Tables III–XII.

[†] The amplification factor is $e^{-\alpha c_i t}$.

Table III. Orr-Summerfeld Equation Eigenvalues ($c_i = -0.008$)

c_r	$Re_{\delta*}$	$\alpha\delta^*$	$(\beta_r v/U_\delta^2) \times 10^6$	$(\beta_i \delta^*/U_\delta) \times 10^3$
0.22462	3893.67	0.08218	4.741	−0.6575
0.25245	2393.42	0.10028	10.577	−0.8022
0.29000	1355.70	0.12946	27.693	−1.0357
0.32000	918.59	0.15778	54.965	−1.2623
0.36000	589.65	0.20561	125.530	−1.6449
0.38000	489.39	0.23607	183.306	−1.8886
0.40000	418.26	0.27431	262.338	−2.1945
0.42000	380.85	0.33298	367.212	−2.6639
0.42409	404.84	0.37051	388.134	−2.9641
0.41573	504.15	0.39671	327.131	−3.1737
0.39705	703.23	0.39816	224.804	−3.1853
0.38037	930.19	0.38702	158.255	−3.0961
0.36625	1178.99	0.37319	115.928	−2.9855
0.34387	1732.63	0.34630	68.729	−2.7704
0.3156	2930	0.3044	32.8	−2.4352

Table IV. Orr-Summerfeld Equation Eigenvalues ($c_i = -0.004$)

c_r	$Re_{\delta*}$	$\alpha\delta^*$	$(\beta_r v/U_\delta^2) \times 10^6$	$(\beta_i \delta^*/U_\delta) \times 10^3$
0.24000	3071.99	0.09537	7.451	−0.3815
0.26000	2194.46	0.11042	13.083	−0.4417
0.29000	1412.88	0.13452	27.610	−0.5381
0.32000	958.80	0.16356	54.589	−0.6542
0.36000	620.48	0.21386	124.082	−0.8554
0.38000	518.97	0.24682	180.723	−0.9873
0.40000	452.97	0.29205	257.902	−1.1682
0.41192	446.16	0.33634	310.528	−1.3454
0.41246	476.61	0.35679	308.772	−1.4272
0.40833	537.70	0.37163	282.217	−1.4865
0.40010	635.18	0.37826	238.266	−1.5130
0.39167	740.10	0.37799	200.041	−1.5120
0.38344	854.32	0.37447	168.073	−1.4979
0.37569	975.79	0.36930	142.178	−1.4772
0.36211	1231.43	0.35731	105.069	−1.4292
0.34021	1801.48	0.33306	62.899	−1.3322
0.32409	2412.16	0.31242	41.976	−1.2497

Table V. Orr-Summerfeld Equation Eigenvalues ($c_i = -0.002$)

c_r	$\mathrm{Re}_{\delta*}$	$\alpha\delta^*$	$(\beta_r \nu/U_\delta^2) \times 10^6$	$(\beta_i \delta^*/U_\delta) \times 10^3$
0.23858	3277.17	0.09607	6.994	−0.1921
0.25400	2487.22	0.10813	11.042	−0.2163
0.27000	1933.16	0.12021	16.890	−0.2404
0.29000	1453.64	0.13687	27.305	−0.2737
0.32000	987.14	0.16656	53.993	−0.3331
0.36000	639.65	0.21809	122.742	−0.4362
0.38000	538.60	0.25307	178.548	−0.5061
0.39400	491.63	0.28592	229.138	−0.5718
0.40346	477.63	0.31405	265.286	−0.6281
0.40665	507.90	0.34456	275.869	−0.6891
0.40399	558.10	0.35811	259.225	−0.7162
0.39664	654.99	0.36656	221.978	−0.7331
0.38856	762.16	0.36738	187.295	−0.7348
0.38064	877.60	0.36463	158.152	−0.7293
0.37319	999.58	0.36016	134.463	−0.7203
0.35979	1260.69	0.34902	99.608	−0.6980

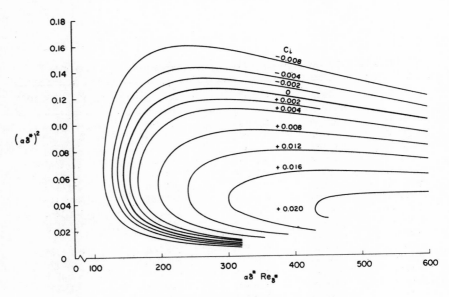

Figure 6.4. Wave Number Squared, $(\alpha\delta^*)^2$,
as a Function of Wave Number Reynolds Number Product,
$\alpha\delta^*$ $\mathrm{Re}_{\delta*}$, for Various Amplification Factors, c_i.

Table VI.   Orr-Summerfeld Equation Eigenvalues ($c_i = 0$)

c_r	$\mathrm{Re}_{\delta*}$	$\alpha\delta^*$	$(\beta_r v/U_\delta^2) \times 10^6$	$(\beta_i \delta^*/U_\delta) \times 10^3$
0.24575	3010.21	0.10265	8.380	$-0-$
0.27001	2033.81	0.12096	16.058	$-0-$
0.29665	1385.97	0.14430	30.885	$-0-$
0.29665	1386.07	0.14429	30.881	$-0-$
0.31813	1039.73	0.16732	51.191	$-0-$
0.33156	897.93	0.18182	67.137	$-0-$
0.34000	817.24	0.19244	80.063	$-0-$
0.36000	667.48	0.22120	119.303	$-0-$
0.36608	627.35	0.23284	135.859	$-0-$
0.38000	563.82	0.25768	173.668	$-0-$
0.38400	544.0	0.270	190.0	$-0-$
0.39089	527.49	0.28436	210.723	$-0-$
0.39500	516.25	0.29769	227.767	$-0-$
0.40029	535.03	0.32709	244.713	$-0-$
0.39908	582.02	0.34363	325.622	$-0-$
0.39628	627.57	0.35056	221.360	$-0-$
0.39287	676.59	0.35424	205.696	$-0-$
0.38523	785.76	0.35634	174.704	$-0-$
0.37763	902.82	0.35444	148.258	$-0-$
0.37040	1026.76	0.35063	126.489	$-0-$
0.35729	1292.40	0.34046	94.122	$-0-$
0.33599	1883.37	0.31858	56.835	$-0-$
0.33052	2078.15	0.31228	49.666	$-0-$
0.31349	2840.17	0.29153	32.178	$-0-$

Table VII.   Orr-Summerfeld Equation Eigenvalues ($c_i = +0.002$)

c_r	$\mathrm{Re}_{\delta*}$	$\alpha\delta^*$	$(\beta_r v/U_\delta^2) \times 10^6$	$(\beta_i \delta^*/U_\delta) \times 10^3$
0.24800	2916.37	0.10797	9.181	0.2159
0.26400	2259.20	0.12012	14.037	0.2402
0.29000	1543.37	0.14221	26.722	0.2844
0.32000	1059.61	0.17144	51.775	0.3429
0.36000	692.20	0.22622	117.652	0.4524
0.38000	590.97	0.26575	170.878	0.5315
0.39231	570.78	0.30659	210.729	0.6132
0.39347	609.85	0.32785	211.523	0.6557
0.38852	703.25	0.34128	188.543	0.6826
0.38167	812.15	0.34476	161.974	0.6895
0.36737	1056.56	0.34073	118.474	0.6815
0.35458	1326.97	0.33158	88.602	0.6632

Table VIII.   Orr-Summerfeld Equation Eigenvalues ($c_i = +0.004$)

c_r	$Re_{\delta*}$	$\alpha\delta^*$	$(\beta_r v/U_\delta^2) \times 10^6$	$(\beta_i \delta^*/U_\delta) \times 10^3$
0.25235	2866.59	0.11163	9.827	0.4465
0.28517	1728.89	0.13881	22.897	0.5553
0.31219	1205.01	0.16597	42.999	0.6639
0.36	720.49	0.23199	115.917	0.9280
0.38	627.71	0.27682	167.581	1.1073
0.38668	645.24	0.30996	185.755	1.2398
0.38366	733.34	0.32727	171.214	1.3091
0.37752	841.96	0.33256	149.112	1.3302
0.37075	961.72	0.33274	128.272	1.3310
0.36405	1089.67	0.33039	110.380	1.3216
0.35163	1364.99	0.32235	83.039	1.2894
0.33101	1978.76	0.30322	50.723	1.2129
0.30867	2986.65	0.27824	28.756	1.1130

Table IX.   Orr-Summerfeld Equation Eigenvalues ($c_i = +0.008$)

c_r	$Re_{\delta*}$	$\alpha\delta^*$	$(\beta_r v/U_\delta^2) \times 10^6$	$(\beta_i \delta^*/U_\delta) \times 10^3$
0.25321	3021.15	0.11717	9.821	0.9374
0.26377	2546.85	0.12564	13.013	1.0052
0.30055	1498.20	0.16020	32.136	1.2816
0.33900	956.55	0.20909	74.100	1.6727
0.36272	785.55	0.25460	117.561	2.0368
0.37118	815.18	0.29441	134.058	2.3553
0.36766	917.55	0.30516	122.277	2.4413
0.35632	1169.05	0.30794	93.860	2.4635
0.34483	1454.43	0.30253	71.727	2.4202
0.32508	2094.79	0.28642	44.449	2.2914
0.30856	2847.65	0.26970	29.223	2.1576

6.11 COMPARISON WITH DATA AND PREVIOUS SOLUTIONS

The projection of the surface of Fig. 6.6 on the $\beta_r v/U^2 - Re_{\delta*}$ plane is plotted in Fig. 6.7 for comparison with the neutral stability data of Schubauer and Skramstad [8]. The minimum critical Reynolds number (minimum $Re_{\delta*}$ on the $c_i = 0$ curve) is found to be 516. A few curves for amplification rates other than zero are shown for reference. The agreement between the calculated neutral curve (heavy line) and the data is seen to

Table X.   Orr-Summerfeld Equation Eigenvalues ($c_i = +0.012$)

c_r	$\mathrm{Re}_{\delta*}$	$\alpha\delta^*$	$(\beta_r \nu/U_\delta^2) \times 10^6$	$(\beta_i \delta^*/U_\delta) \times 10^3$
0.25816	2999.70	0.12701	10.931	1.5242
0.27789	2213.05	0.14460	18.157	1.7352
0.29628	1712.89	0.16346	28.275	1.9616
0.3200	1284.29	0.19299	48.087	2.3159
0.3400	1065.06	0.22653	72.316	2.7184
0.35341	1036.93	0.27003	92.032	3.2403
0.35080	1147.04	0.27899	85.322	3.3479
0.34591	1290.13	0.28183	75.564	3.3819
0.33635	1571.66	0.37996	59.913	3.3595
0.31784	2242.36	0.26757	37.926	3.2108
0.30238	3005.65	0.25352	25.505	3.0423

Table XI.   Orr-Summerfeld Equation Eigenvalues ($c_i = +0.016$)

c_r	$\mathrm{Re}_{\delta*}$	$\alpha\delta^*$	$(\beta_r \nu/U_\delta^2) \times 10^6$	$(\beta_i \delta^*/U_\delta) \times 10^3$
0.26001	3175.78	0.13603	11.137	2.1765
0.30082	1782.62	0.17951	30.292	2.8721
0.3200	1447.61	0.20852	46.095	3.3364
0.33040	1369.51	0.23367	56.374	3.7387
0.33127	1459.05	0.24674	56.020	3.9478
0.32479	1744.02	0.25229	47.985	4.0367
0.30849	2445.33	0.24537	30.954	3.9259
0.29848	2976.62	0.23785	23.851	3.8057

Table XII.   Orr-Summerfeld Equation Eigenvalues ($c_i = +0.020$)

c_r	$\mathrm{Re}_{\delta*}$	$\alpha\delta^*$	$(\beta_r \nu/U_\delta^2) \times 10^6$	$(\beta_i \delta^*/U_\delta) \times 10^3$
0.28000	2666.62	0.16836	17.678	3.3672
0.29000	2361.07	0.18221	22.380	3.6442
0.29797	2185.80	0.19572	26.680	3.9144
0.30131	2156.63	0.20402	28.505	4.0805
0.30197	2285.94	0.21365	28.223	4.2731
0.30160	2317.91	0.21433	27.888	4.2866
0.29455	2782.43	0.21564	22.828	4.3128

be excellent except at the lowest Reynolds number and highest frequencies, where a small error in c_i corresponds to a large error in frequency. It should also be noted that the scatter in the data appears to be greatest near the minimum Reynolds number.

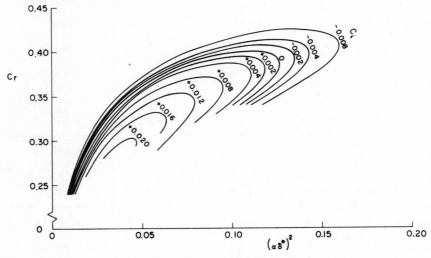

Figure 6.5. Propagation Velocity, c_r, as a Function of Wave Number Squared, $(\alpha \delta^*)^2$, for Various Amplification Factors, c_i.

The $\alpha\delta^*$ versus Re_δ^* projection of the surface of Fig. 6.6 is also plotted, in Fig. 6.8, for comparison with the data of Schubauer and Skramstad, and for comparison with other solutions to the Orr-Sommerfeld equation [Eq. (6.2)]. Again, there is good agreement between the calculations made by the method of this chapter and the data, except near the maximum wave number and minimum Reynolds number. All theoretical solutions agree well on the lower branch, while on the upper branch Schlichting's solution [7] is low and Shen's [10] is high. The solution of Kurtz and Crandall [11] is slightly above the present solution over a small range. Kaplan's solution [12] coincides with the present solution for the scale shown.

Discrepancies in the otherwise excellent agreement between the data and the present theory in the neighborhood of the minimum Reynolds number, together with the apparent scatter in the data in that region, indicates that perhaps the data may be somewhat incorrect. The curve for $c_i = -0.008$ shows that a small uncertainty in measuring the amplification rate in this

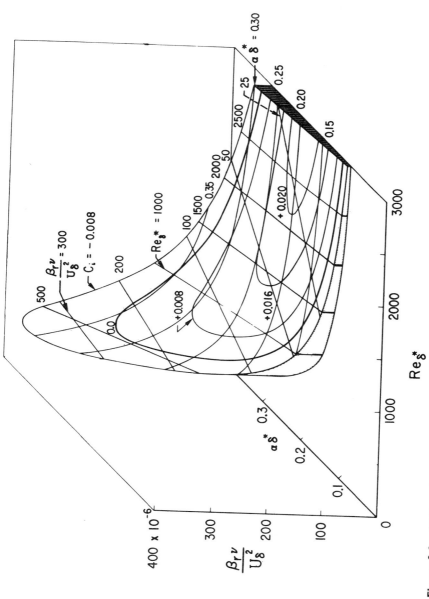

Figure 6.6. Oblique Projection of Frequency, $\beta_r \nu / U_\delta^2$, as a Function of Wave Number, $\alpha\delta^*$, and Reynolds Number, Re_{δ^*}, Showing Curves of Constant Amplification Factor c_i, and the Traces of Planes of Constant Frequency, Wave Number, and Reynolds Number.

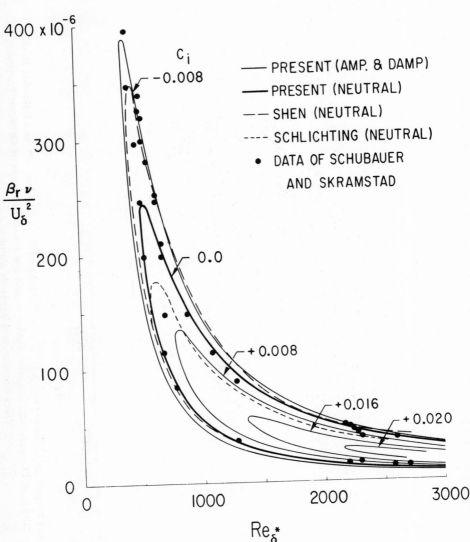

Figure 6.7. Comparison of the $\beta_r \nu / U_\delta^2$ — $Re_{\delta*}$ Projection of the Amplification Surface With Neutral Stability Data of Schubauer and Skramstad and the Theories of Schlichting and Shen.

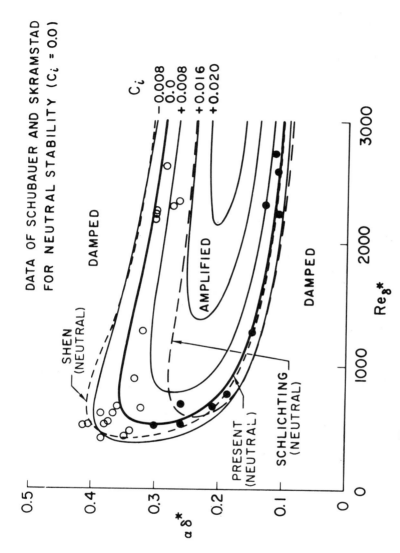

Figure 6.8. Comparison of the $\alpha\delta^*$ — Re_{δ^*} Projection of the Amplification Surface With Neutral Stability Data of Schubauer and Skramstad and the Theories of Schlichting and Shen.

region would lead to large uncertainties in both wave number and Reynolds number.

An alternative representation of the neutral and neighboring amplification curves, in terms of c_r and Re_δ^*, is shown in Fig. 6.9. The measurements of Schubauer and Skramstad [8] and the solution of Schlichting [7] are also indicated. Again, the greatest discrepancy between the computed neutral curve and the data occurs near the minimum Reynolds number. The experimental data tend to cover an area in this region rather than to define a curve. The solution of Schlichting tends to fall above the data in most areas, except at large Reynolds numbers on the lower branch.

Figure 6.10 compares the dimensionless amplification exponent $\beta_i \delta^*/U_\delta$ with the data of Schubauer and Skramstad [8] and the solutions of Schlichting [7] and Shen [10]. There is good agreement between the present solution and the data, except at the lowest Reynolds number where the present solution still compares favorably with the other solutions. Unlike the asymptotic solutions of Schlichting [7] and Shen [10], the present solution extends into the damping region $(\beta_i \delta^*/U_\delta < 0)$, where agreement with the data is as good as at positive values.

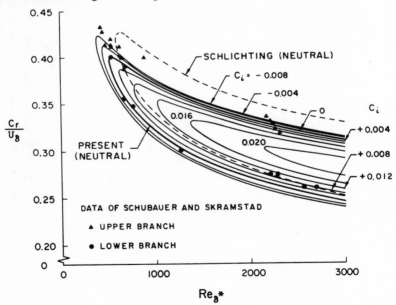

Figure 6.9. Comparison of the c_r/U_δ — Re_{δ^*} Representation of the Amplification Surface With Neutral Stability Data of Schubauer and Skramstad and the Theory of Schlichting.

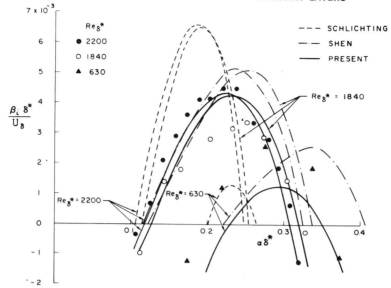

Figure 6.10.   Comparison of the $\beta_i \delta^*/U_\delta$ Representation of the Amplification Surface With Amplification Data of Schubauer and Skramstad and the Theories of Schlichting and Shen.

Root-mean-square (rms) profiles of perturbation velocity in the direction of flow, u'_{rms}/U_δ, are compared in Fig. 6.11 with the measurements of Schubauer and Skramstad [8]. The data were taken with a hot wire anemometer that has an output proportional to the rms value of the velocity measured, so that the phase reversal as a function of y/δ should appear as a sharp angle. Since the real and imaginary parts of $d\phi/dy$ do not pass through zero at the same value of y/δ, the ϕ_{rms} curve does not actually reach zero at the point of discontinuous slope, although this fact cannot be seen at the scale to which the figure is plotted.

The amplitude of the perturbation in the linearized theory is arbitrary; for this reason, the theoretical u'_{rms} curve must be multiplied by a scale factor, for comparison with data. Because the scatter in the data is small at the second maximum (near $y/\delta = 1$), the ordinates of the data and the computed solution were made to coincide at that maximum of each set of data. The abscissas of the computed solutions were not scaled.

While the agreement between the adjusted theory and experiment is excellent for both upper and lower branches beyond the phase reversal points, the data appear to be fitted better by the upper-branch solution

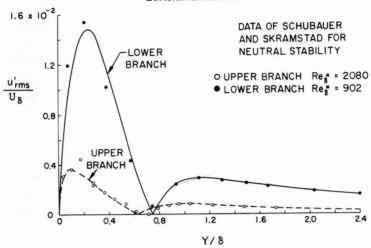

**Figure 6.11. Comparison of the
Root-Mean-Square Longitudinal Disturbance Velocity,
u'_{rms}/U_δ, as a Function of Wall Distance, y/δ,
with Data of Schubauer and Skramstad.**

than by the lower-branch solution between the reversal point and the wall ($y/\delta = 0$). Since the absolute magnitude of the lower-branch perturbation velocity is greater than 1% of the undisturbed free-stream velocity, it is possible that nonlinear effects may already have become noticeable in the oscillation at this amplitude.

Although the limit of the linearized theory may be at most a few percent, in the sections below we shall present some plots of properties of much larger disturbances, in order to illustrate the nature of the disturbed flows.

6.12 STREAM FUNCTIONS OF THE DISTURBED FLOW

The stream function of the perturbation on the boundary layer ψ' is given by Eq. (6.1) and represents a Tollmien-Schlichting wave. The total stream function ψ is

$$\psi = \Psi(y) + C_s\psi'(x, y, t) \tag{6.12}$$

where $\Psi(y)$ is the y-dependent part of the basic-flow stream function. It should be noted that the x-dependent factor of Ψ has been discarded by the original assumption of parallel flow. The arbitrary amplitude factor C is discussed later in this section.

In our example, we have taken the basic flow to be the incompressible laminar boundary layer on a flat plate with zero pressure gradient. Thus, for this case, $\Psi(y)$ is the Blasius function (see Chapter 3 for the computation of a generalized Blasius function). In general, however, the method could be applied to any boundary-layer flow, similar or dissimilar, for which the parallel flow approximation is acceptable.

Because $\psi(x, y, t)$ is a function of three variables, it is convenient to fix the value of either x or t, which are equivalent on the neutral amplification curve. Fixing x and plotting $\psi(y, t)$ corresponds to observing the disturbance as it passes a point. Fixing t and plotting $\psi(x, y)$ corresponds to taking a snapshot of a portion of the flow. The latter approach has been taken in the stream-function plots presented in this section and in the velocity plots presented in the next section. It should be noted that ψ', hence ψ, is periodic in x since α is real; therefore, only one wave length need be displayed.

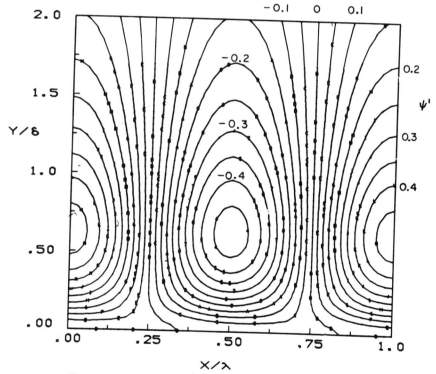

Figure 6.12. Perturbation Stream Function at $t = 0$
for Upper Branch of Neutral Stability Curve.

Due to the linearization in the derivation of the Orr-Sommerfeld equation, the amplitude of ψ' is arbitrary. In the stream-function plots in this chapter and in Chapter 7, the amplitude factor C_s in Eq. (6.12) is chosen so that

$$C_s = \max_{y} (u'(y))/U_\delta \qquad (6.13)$$

is a specified fraction.

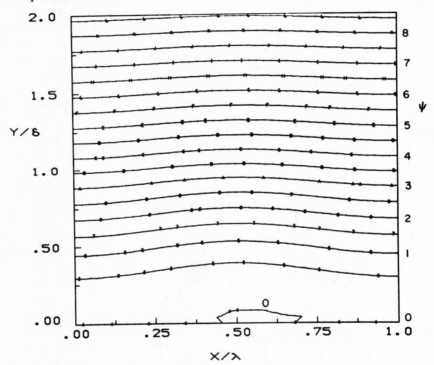

Figure 6.13.   Total Stream Function at $t = 0$ for $u'_{\max}/U_\delta = 0.10$
for Upper Branch of Neutral Stability Curve.

The series of stream-function plots, Figs. 6.12 through 6.15, begins with the perturbation stream function ψ' and the total stream function for $C_s = 0.1, 0.5$ are next shown. The last of this series, Fig. 6.15, is plotted with $\delta = 2$, so that x and y are to the same scale. The series is for $\mathrm{Re}_{\delta*} = 2080$ on the upper branch of the neutral amplification curve, which corresponds to data of Schubauer and Skramstad [8].

An eddy can clearly be identified in Fig. 6.15. Although the linearized theory can hardly be expected to be applicable at 50% disturbance, the figure clearly indicates the nature of the Tollmien-Schlichting wave, which is not readily apparent at smaller amplitudes. The asymmetry near the wall is due to the contribution of the imaginary part of the eigenfunction $\phi_i(y)$, which is vital for a sustained or growing disturbance.

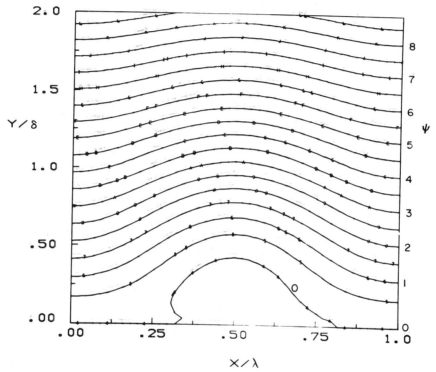

**Figure 6.14. Total Function at $t = 0$ for $u'_{max}/U_\delta = 0.50$
for Upper Branch of Neutral Stability Curve.**

While the stream function plots of Figs. 6.11 through 6.15 represent the unsteady instantaneous view of the flow moving past the observer, fixed with respect to the wall, the steady state flow pattern can be obtained by subtraction of the wave velocity from the velocity field. Figure 6.16 is the steady flow counterpart of Fig. 6.15 as seen by an observer moving with the wave speed. The dashed line in the figure is the locus of points with zero horizontal velocity in the moving reference frame. The eddies now represent the classical "cat's eye" pattern (6.21), which appears more rounded as a result of the present more exact calculations.

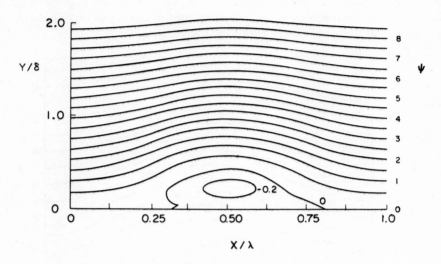

Figure 6.15. Total Stream Function with x **and** y **to Same Scale at** $t = 0$ **for** $u'_{max}/U_\delta = 0.50$ **for Upper Branch of Neutral Stability Curve.**

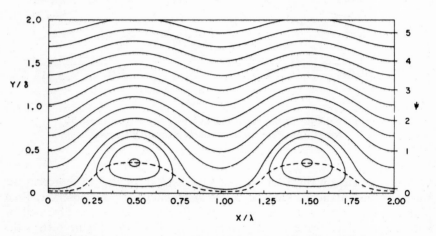

Figure 6.16. Stream Function in Reference Frame Moving with Propagation Velocity of Disturbance "Cat's Eye".

6.13 VELOCITY OF THE DISTURBED FLOW

The components of the perturbation velocity are given by

$$u'(x, y, t) = \text{real}\,\{d\phi/dy \, \exp(i(\alpha x - \beta t))\}$$
$$v'(x, y, t) = \text{real}\,\{i\alpha\phi \quad \exp(i(\alpha x - \beta t))\}$$

$$(6.14)$$

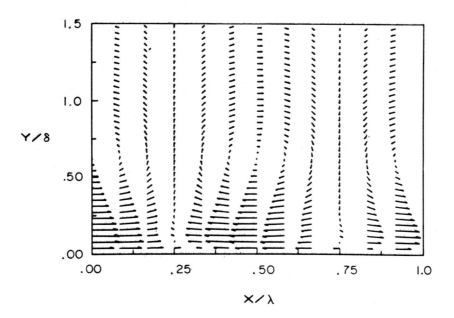

**Figure 6.17. Perturbation Velocity at $t = 0$ for Upper Branch
of Neutral Stability Curve.**

Furthermore, the components of the total velocity are

$$u = U + u' = df(y)/dy + u'(x, y, t)$$
$$v = V + v' = 0 + v'(x, y, t)$$

$$(6.15)$$

It should be remembered that a parallel basic flow has been assumed.

Two vector plots of perturbation, as well as total velocity, are shown in Fig. 6.17 and Fig. 6.18, respectively, corresponding to $\text{Re}_{\delta*} = 2080$ on the upper branch of the neutral curve. The scale of the vectors was determined by the requirement that neighboring vectors not overlap. The relative magnitudes of the vectors and the angles are true to scale, but the vertical scale of the grid has been stretched to show more clearly the disturbance.

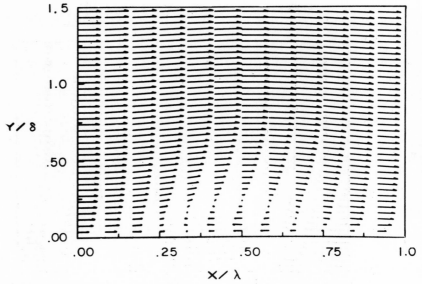

Figure 6.18.   Total Velocity at $t = 0$ for $u'_{max}/U_\delta = 0.50$ for Upper Branch of Neutral Stability Curve.

These figures, as well as those of the previous section, were computed on the IBM 7094 and plotted on the SC 4020 [22].

6.14 PERTURBATION PRESSURE

The partial derivatives of the perturbation pressure p' are given by the perturbation equations of motion, which are derived from the Navier-Stokes equations. At large distances from the wall (i.e., $y \to \infty$), the perturbation pressure $p'(x, \infty)$ must approach zero. The normal pressure derivative may be integrated from infinity toward the wall at any x to give the whole pressure field; thus, only the y equation of motion is required. Upon integration and substitution of the solution for large y, Eq. (6.6), there results

$$p'(x, y, t) = -\alpha \exp\left[i(\alpha x - \beta t)\right]\left\{\frac{i}{\mathrm{Re}}\frac{d\phi(y)}{dy}\int_y^{y_0}\left[\frac{i\alpha^2}{\mathrm{Re}} + \beta - U(y)\right]\phi(y)\,dy\right.$$

$$\left. + \left[\frac{i\alpha^2}{\mathrm{Re}} + \beta - \alpha\right]\left[-\frac{A}{f}\exp(+fy_0) + \frac{1}{\alpha}\exp(-\alpha y_0)\right]\right\} \qquad (6.16)$$

Only the real part of p' is significant. The integral in this expression for p' requires a quadrature of the results of the quasilinearization program that is carried out by Simpson's rule. The solution in this chapter is for a flat plate without pressure gradient, so that

$$p = p_0 + p' \qquad (6.17)$$

where p_0 is a positive constant representing the external pressure level. Negative values of p' are admissible if p remains positive.

Figure 6.19 shows the perturbation pressure for $\mathrm{Re}_{\delta*} = 2080$ on the upper branch of the neutral curve. Perturbation pressure in dimensional units,

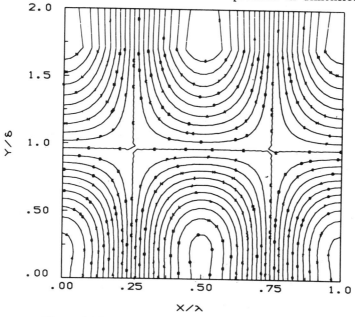

Figure 6.19. Perturbation Pressure for $\mathrm{Re}_{\delta*} = 2080$ on the Upper Branch of the Neutral Stability Curve.

for a proportion of disturbance C_s, is obtained by multiplying p' by $C_s \rho U_\delta^2 / (d\phi/dy)_{max}$, where ρ and U_δ are now in dimensional units.

6.15 POWER INPUT TO THE DISTURBANCE

The power that feeds the disturbance in the laminar boundary layer has its origin in the work done by the basic flow against the Reynolds stresses of the disturbance. In incompressible flow, mechanical energy and pressure work are decoupled from other forms of energy. Thus, the energy per unit time and volume feeding the disturbance may be derived from the equations of motion by multiplying by the respective velocities, u and v.

A more useful quantity than the increase of kinetic energy due to the disturbance is the square of the perturbation velocity. This may be thought of as related to the kinetic energy which would be measured in a reference frame moving with the local velocity of the basic flow. The components of the normalized perturbation kinetic energy u' and v' can be derived from the perturbation equations of motion by multiplying by u' and v', respectively, to obtain

$$\frac{D(u'^2/2)}{Dt} + u'\frac{\partial p'}{\partial x} - \frac{1}{Re}\nabla^2(u'^2/2) = -u'v'\frac{dU}{dy} \tag{6.16}$$

$$\frac{D(v'^2/2)}{Dt} - \frac{1}{Re}\nabla^2(v'^2/2) = -v'\frac{\partial p'}{\partial y} \tag{6.17}$$

where D/Dt is the "substantial" derivative.

The first two terms in Eq. (6.16) represent rates of energy storage in, or retrieval from, kinetic energy or pressure fields. The third term represents viscous dissipation—a sink. The right-hand side represents the rate of exchange of energy with the basic flow by means of the Reynolds stress $(u'v')$, which is a net source of energy over a whole oscillation cycle. The first and last terms in Eq. (6.17) similarly represent rates of exchange of energy with kinetic or pressure fields. Dissipation is represented by the second term. The pressure term in Eq. (6.17) has been placed on the right, because, on balance, it plays the part of the source. Thus, energy is fed from the basic flow to the x component of the oscillation, which in turn feeds by way of the pressure field to the y component of the oscillation.

Clearly, the amplification or damping of the disturbance depends on the balance between the power P taken from the basic flow $u'v'\,dU/dy$ and the dissipation $\nabla^2(u'^2 + v'^2)/(2\,Re)$. Although the dissipation is independent of the phases of u' and v', the phase difference between u' and v' in the source term is important. In order to demonstrate the importance of the phase, u' and v' are rewritten in the form

$$u' = \text{real} \left[\left| \frac{d\phi}{dy} \right| \exp\left[i(\theta + \angle d\phi/dy) \right] \right] \tag{6.18}$$

$$v' = \text{real} \left[\left| \phi \right| \exp\left[i(\theta + \angle \phi + \pi/2) \right] \right] \tag{6.19}$$

where $\angle \phi$ denotes the phase angle of ϕ.

The power P per unit volume becomes

$$P = -\tfrac{1}{2} |\phi| \left| \frac{d\phi}{dy} \right| \frac{dU}{dy} \left[\sin(2\theta + \phi + \angle d\phi/dy) + \sin(\angle \phi - \angle d\phi/dy) \right] \tag{6.20}$$

The power P into the disturbance over one wave length is

$$P = \tfrac{1}{2} \int_0^2 \int_0^\infty P \, dy \, dx = \pi \int_0^\infty |\phi| \left| \frac{d\phi}{dy} \right| \frac{dU}{dy} \sin(\angle \phi - \angle d\phi/dy) \, dy \tag{6.21}$$

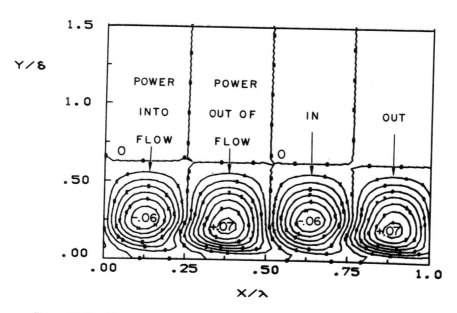

Figure 6.20.   Power to Tollmien-Schlichting Wave From Basic Flow, $-u'v'dU/dy$, for Upper Branch of Neutral Curve.

If ϕ has only a real part, then $\angle\phi = \angle d\phi/dy$ and $P_\lambda = 0$. Thus, the imaginary part is necessary for energy transfer to occur.

The power per unit volume to the basic flow $u'v'\, dU/dy$ is plotted in Fig. 6.20 for $\mathrm{Re}_\delta^* = 2080$ again on the upper branch of the neutral stability curve, where the region $0 \leqslant y/\delta \leqslant 0.7$ is divided into four parts, alternatively positive and negative. Power into the disturbance is represented by the positive regions. Although this plot represents a neutrally stable solution, there is a net power transfer to the disturbance to supply the power lost by viscous dissipation.

6.16 CONCLUSION

The mathematical techniques applied in this chapter differ more in degree than in kind from those applied in Chapter 5. The scale of the problem was larger because four homogeneous solutions were required compared to three in Chapter 5 and the dimension of the state vector was increased to 12. However, this increase in size caused no problem apart from an increase in computation time.

Use of an analytic solution to a form of the differential equations approached at large independent variable values as a boundary condition has already been seen in Chapter 5. However, in the present chapter the solution was made without first transforming to a finite interval. Also, all the initial values were derived from the approximate analytical solution.

We have seen that in a problem with homogeneous boundary conditions that depends on a number of parameters, there can be a choice of which parameters are to be specified and which parameters are to be determined as eigenvalues in the course of the solution. The particular choice will depend on the region of parameter space in which the search for a solution is being conducted.

Finally we have found restriction of convergence during early iterations to be a useful tool. This procedure will be employed again in Chapter 7, and in Chapter 8, it will be found to be critical for convergence.

REFERENCES

1. L. M. Mack, Computation of the stability of the laminar compressible boundary layer *in* "Methods in Computational Physics," Vol. 4, pp. 247–299. Academic Press, New York, 1965.
2. W. H. Reid, The stability of parallel flows in basic developments *in* "Fluid Dynamics," Vol. 1, pp. 249–307. Academic Press, New York, 1965.

3. W. Eckhaus, "Studies in Non-Linear Stability Theory," p. 8. Springer-Verlag, Berlin and Heidelberg, 1965.
4. O. Reynolds, An experimental investigation of the circumstances which determine whether the motion of water shall be direct or sinuous, and of the law of resistance in parallel channels in "Scientific Papers," Vol. 2, pp. 51–105. Cambridge Univ. Press, New York and London, 1883.
5. W. Tollmien, General Stability Criterion of Laminar Velocity Distributions", NACA Tech. Memo 792 (1936).
6. H. Schlichting, Zur Entstehung der Turbulenz bei der Plattenströmung. *Nachr. Ges. Wiss. Göttingen, Math. Phys. Kl.* 181–208 (1933); also *Z. angew. Math. Mech.* **13**, 171–174 (1933).
7. H. Schlichting, Amplitudenverteilung und Energiebilanz der kleinen Störungen bei der Plattenströmung. *Nachr. Ges. Wiss. Göttingen, Math. Phys. Kl., Fachgruppe I*, **1**, 47–78 (1935).
8. G. B. Schubauer and H. K. Skramstad, "Laminar Boundary-Layer Oscillations and Transition on a Flat Plate," Nat. Bur. of Standards Res. Paper 1172; also Laminar boundary-layer oscillations and stability of laminar flow. *J. Aeron. Sci.* **14**, 69 (1947).
9. C. C. Lin, "The Theory of Hydrodynamic Stability," Cambridge Univ. Press; Cambridge, Massachusetts, 1955.
10. S. F. Shen, Calculated amplified oscillations in plane Poiseuille and Blasius flows. *J. Aeron. Sci.* **21**, 62–64 (1954).
11. E. F. Kurtz, Jr. and S. H. Crandall, Computer aided analysis of hydrodynamic stability. *J. Math. and Phys.*, **41**, 264–279 (1962).
12. R. E. Kaplan, "The Stability of Laminar Incompressible Boundary Layers in the Presence of Compliant Boundaries," ASRL TR 116-1 (1964).
13. P. R. Nachtsheim, "Stability of Free-Convection Boundary Layer Flows." NASA TND-2089 (1963).
14. J. R. Radbill and E. R. van Driest, "A New Method for Prediction of Stability of Laminar Boundary Layers," AFOSR 66-0702 (1966).
15. H. Schlichting, "Boundary Layer Theory," pp. 314–316. McGraw-Hill, New York, 1959.
16. S. Pai, "Viscous Flow Theory I—Laminar Flow," pp. 309–311. Van Nostrand, Princeton, New Jersey, 1956.
17. F. B. Hildebrand, "Introduction to Numerical Analysis," Chapter 4. McGraw-Hill, New York, 1956.
18. S. A. Gill, A process for the step-by-step integration of differential equations in an automatic digital computing machine. *Proc. Cambridge Phil. Soc.* **47**, 96–108 (1951).
19. F. B. Hildebrand, "Introduction to Numerical Analysis," Chapter 6, p. 197. McGraw-Hill, New York, 1956.
20. G. A. McCue, M. M. Dworetsky, and H. Du Prie, "FORTAN IV Stereographic Function Representation and Contouring Program," NAA S&ID, S&ID 65-1182 (1965).

Chapter 7

PREDICTION OF THE STABILITY
OF LAMINAR PIPE FLOW

7.1 INTRODUCTION

A characteristic value problem containing a very large parameter is presented in this chapter, for several "modes" of the multiple valued solution are obtained. This final fluid mechanics problem of the present volume is a variation of the laminar flow stability problem of Chapter 6, wherein the so-called Poiseuille flow in a two-dimensional duct is examined. A distinguishing feature of the two-dimensional Poiseuille flow stability problem is the much higher Reynolds number at which self-sustaining disturbances first occur. The large magnitude of this parameter in the differential equation accelerates the growth of unwanted solutions to the point where the orthogonalization procedure employed in earlier chapters is no longer adequate, and an additional device must be utilized to obtain a solution.

A transformation that removes the explicit dependence of the equation on the large parameter is accomplished by stretching the independent variable. The elements in the Jacobian matrix become, at most, of order unity, and the need for orthogonalization is reduced or eliminated. In spite of the increase in the length of the integration interval, there is a net saving in the computation required to obtain a solution.

Because of the symmetry of the two-dimensional Poiseuille flow, odd and even eigenfunctions can be separated by prescribing a boundary condition at the center line of the channel.

Estimates of the form of higher order eigenfunctions are easier to obtain for this geometry than for that in Chapter 6. As a result of these considerations, it has been possible to obtain the second-, third-, and fourth-order eigenfunctions and eigenvalues for the two-dimensional Poiseuille flow instability.

7.2 PHYSICAL BACKGROUND

Fully developed laminar flow in a two-dimensional channel or pipe is the limit of a very wide and thin duct and is seldom approached in practice. However, the two-dimensional disturbances in this geometry are easier to handle mathematically than the three-dimensional disturbances found in a circular pipe. The two-dimensional channel flow is a truly parallel flow, in contrast to the boundary-layer flow on a flat plate, which it resembles.

As was the case with the boundary layer, the channel flow becomes unstable to traveling wave disturbances as the Reynolds number increases, and at even larger Reynolds numbers, the laminar flow breaks down into turbulent flow. The Reynolds number for the channel flow is evaluated in terms of the total wall spacing. When the minimum-critical Reynolds number, for instability based on half the channel width, is compared with the minimum-critical Reynolds number, for the flat-plate boundary layer based on the boundary-layer thickness, it is found to be several times as large. This is probably due to the stabilizing presence of the second wall.

7.3 EIGENVALUE PROBLEM FOR TWO-DIMENSIONAL POISEUILLE FLOW

The eigenvalue problem for the two-dimensional Poiseuille flow is formulated in terms of Eq. (6.3), the Orr-Sommerfeld equation, which we reproduce here for reference:

$$(U-c)(\phi''-\alpha^2\phi)-U''\phi = -(i/\alpha \mathrm{Re}_D)(\phi^{iv}-2\alpha^2\phi''+\alpha^4\phi) \qquad (7.1)$$

Primes denote differentiation with respect to the independent variable y, which runs across the channel. Unlike the problem in Chapter 6, the homogeneous boundary conditions are imposed at the ends of a finite interval and are

$$y = 0, \quad \phi = \phi' = 0; \quad y = 1, \quad \phi = \phi' = 0 \qquad (7.2)$$

Also, unlike the problem in Chapter 6, the velocity of the undisturbed flow $U(y)$ can be expressed in a simple formula,

$$U(y) = 4y(1-y) \qquad (7.3)$$

rather than as a tabulated function or the solution of a differential equation. The Reynolds number for this problem is defined by

$$\mathrm{Re}_D = U_{max} D/\nu$$

where U_{max} is the maximum velocity of the undisturbed flow, which occurs at the center of the channel, D is the channel width, and ν is the kinematic viscosity. Refering to the assumed dimensionless form of the stream function, for the traveling wave perturbation on the basic parallel flow, Eq. (6.1), in the form

$$\psi = \phi \exp\left[i\alpha(x-ct)\right] \tag{7.4}$$

shows that $\phi(y)$ is the y-dependent part of the perturbation stream function, where α is the wave number and c is a complex propagation velocity, the imaginary part of which governs the growth of the disturbance.

When two of the real parameters c_r, c_i, α, Re_D are specified, the remaining two must be determined as eigenvalues, so that a solution of Eq. (7.1), satisfying Eq. (7.8), may be found. Before the quasilinearization method is applied to this problem, we shall transform the equation to make it more stable toward the unwanted parasitic solutions, which grow rapidly because of the presence of the large parameter αRe_D.

7.4 TRANSFORMATION OF THE INDEPENDENT VARIABLE

The Orr-Sommerfeld equation (7.1) can be written in the form

$$\phi^{iv} = i\alpha Re_D((U-c)(\phi''-\alpha^2\phi)-U''\phi)+2\alpha^2\phi''-\alpha^4\phi \tag{7.5}$$

to show the effect of αRe_D, which is a very large number for cases of interest in Poiseuille flow, e.g., 20,000 (see Nachtsheim [1]). This large factor will amplify small errors in the terms in parentheses and cause large errors in the fourth derivative, which will, of course, lead to large errors in the lower derivatives. Another way of looking at this is that the Jacobian matrix, shown in Section 6.5, contains three very large elements in its first row that dominate the others.

We wish to find a transformation that removes the explicit dependence of Eq. (7.5) on the large parameter αRe_D and that makes the largest element in the Jacobian matrix of order one. The desired result is obtained by introducing a new "stretched" independent variable η, defined by

$$\eta = y(\alpha Re_D)^{1/2} \tag{7.6}$$

Expansions in $(i\alpha Re)^{-1/2}$ have been used in WKB solutions of the Orr-Sommerfeld equation (see Reid [2]).

The result of substituting Eq. (7.6) into Eq. (7.5) is

$$\phi^{iv} = (2(\alpha/Re_D)+i(U-c))\phi''-((\alpha/Re_D)(i(U-c)+(\alpha/Re_D))+iU'')\phi \tag{7.7}$$

with the boundary conditions

$$\eta = 0, \quad \phi = \phi' = 0; \quad \eta = (\alpha Re_D)^{1/2}, \quad \phi = \phi' = 0 \tag{7.8}$$

The factor αRe_D does not appear in the transformed differential equation, although it does appear in the boundary conditions, and the parameter α^2 is replaced by α/Re_D.

We note that, as α/Re_D approaches zero, Eq. (7.7) approaches the limiting form

$$\phi^{iv} = i((U-c)\phi'' + U''\phi) \tag{7.9}$$

which depends on α and Re_D only through the length of the integration interval $(\alpha \text{Re}_D)^{1/2}$. Since U is of order one and ϕ can be chosen of order one, the terms on the right side of Eq. (7.9) may be expected to be of order $(\alpha \text{Re}_D)^{-1}$, which may still be finite. We shall find (Section 7.7) that the wave frequency becomes rapidly less dependent on α as Re_D increases, which is consistent with Eq. (7.9).

7.5 JACOBIAN MATRIX

In the solution of Eq. (7.7) by quasilinearization, several choices of eigenvalues are possible, as was the case with the untransformed equation (6.3). Either αRe_D and c_i can be fixed, and α/Re_D and c_r sought as eigenvalues; or Re_D and α/Re_D can be fixed and c_r and c_i determined as eigenvalues. The use of Re_D as an eigenvalue does not appear attractive because this would result in an integration interval of varying length. Columns for the use of c (complex or c_r) and α/Re_D are shown in the Jacobian matrix below. The quantity that is differentiated, X'_i, is shown in the left column and the variable X_j, by which X'_i is differentiated, is shown in the first row: (see formula on page 116).

The functions U and U'' are calculated according to

$$U = U(y(\eta)), \qquad U'' = (1/\alpha \text{Re}_D)\,(d^2 U/dy^2) = -8/(\alpha \text{Re}_D)$$

7.6 SELECTION OF HIGHER MODES

Any function on a finite interval can be represented as the sum of even and odd functions. By solving for the even and odd parts of the solution separately, it is possible to impose even or odd boundary conditions at the center line of the channel and seek the solution over only half the range. That is, for the even part (first, third, ... modes), the odd derivatives are zero at the center line, and for the odd part (second, fourth, ... modes), the function and even derivatives are zero at the center line. By imposing the boundary condition that even or odd derivatives be zero at the center

$$
\begin{array}{c|cccccc}
\overset{X_i'}{\downarrow} \quad X_j \to & \phi''' & \phi'' & \phi' & \phi & c & \dfrac{\alpha}{\text{Re}_D} \\
\hline
(\phi''')' & 0 & \dfrac{2\alpha}{\text{Re}_D}+i(U-c) & 0 & \left\{-\dfrac{\alpha}{\text{Re}_D}\left[i(U-c)+\dfrac{\alpha}{\text{Re}_D}\right]+iU''\right\} & -i\left[\phi''-\dfrac{\alpha}{\text{Re}_D}\phi\right] & 2\left[\phi''-\dfrac{\alpha}{\text{Re}_D}\phi\right]-i(U-c)\phi \\
(\phi'')' & 1 & 0 & 0 & 0 & 0 & 0 \\
(\phi')' & 0 & 1 & 0 & 0 & 0 & 0 \\
\phi' & 0 & 0 & 1 & 0 & 0 & 0
\end{array}
$$

line $(\eta = (\alpha \mathrm{Re}_D)^{1/2}/2)$, odd or even modes, respectively, are obtained, which doubles the separation between solutions to which the algorithm can converge.

Selection of higher modes from among those of prescribed parity is accomplished by specifying the initial approximations for the eigenvalues and by restricting the change of these estimates on early iterations. In this manner, the eigenfunction assumes the form consistent with the estimated eigenvalues before the eigenvalues are allowed to change very much. The number of iterations required for this to occur will depend on how good the initial estimate of the eigenfunction is. The number of cycles that are restricted, accordingly, may be varied using *a priori* knowledge of the quality of the estimate. For example, no restriction is necessary if a solution for a closely neighboring point in parameter space is available, whereas several cycles of restricted search would be specified if the first point of a new mode is sought. Eigenvalues for higher modes are usually easier to estimate than eigenfunctions, so that restricting the initial change of the value rather than the function appears to be the logical approach.

The restriction of the changes in the eigenvalues during the initial iterations, where they are likely to be large, is justifiable from another point of view. In regions of parameter space where the curvature is large or where the initial estimate is far from the true solution, the linear extrapolations furnished by the quasilinearization algorithm may yield a new approximation, which is closer to another mode of the solution. However, the direction of the change is usually much better than the magnitude, so that convergence can be secured by taking a fraction of the indicated step.

A simple but effective method for implementing the extension of a solution has been to plot curves of some constant parameter (e.g., c_i) on graph paper and to extrapolate them with a French curve. The solutions are obtained sequentially by using the graphically obtained eigenvalue initial estimates and by using the eigenfunctions obtained for each point as initial estimates for the next point. On the basis of the very limited number of higher mode solutions obtained, the mode surfaces in parameter space appear to be well separated on c_r. Thus, once one point is located on a mode, others can be found with little difficulty.

7.7 EIGENFUNCTIONS AND EIGENVALUES

The results of the two-dimensional Poiseuille flow computations are given in this and the following section. The eigenfunctions and eigenvalues are shown first. These are followed by stream function and vector velocity plots and plots of the power transferred from the basic flow to the disturbance.

A comparison of eigenfunctions for the first four modes is shown in Fig. 7.1. The eigenfunctions are normalized by dividing each mode by the maximum value of the real or imaginary part for that mode. Since the phase, as well as the magnitude of the solutions to the complex linear equation, are arbitrary, the real part is set nonzero (later normalized as

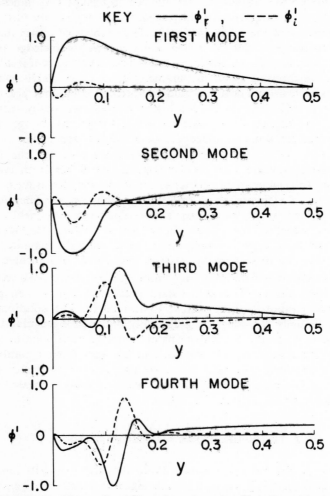

Figure 7.1 Normalized Eigenfunctions for First Four Modes.

above) and the imaginary part set zero at the center line for even modes. The first derivative is set in the same manner for odd modes. In general the number of zero crossings of the real part of the function increases by one for each mode number. The relative size of the imaginary part is small for the first mode, but the imaginary part increases in relative size with increasing mode number. Due to the increase of c_r, the real part of the propagation velocity, with increasing mode number, the largest magnitude of the velocity (i.e., ϕ') occurs farther and farther from the walls ($y = 0, D$). This velocity also becomes increasingly, sharply peeked near its maxima.

Eigenvalues are shown in Figs. 7.2 and 7.3 on the basis actually used in the computations: $\alpha D/\mathrm{Re}_D$ as a function of $\alpha D \mathrm{Re}_D$, and c_r as a function of $\alpha D \mathrm{Re}_D$, both with c_i parameters. Curves on the "amplification surface"

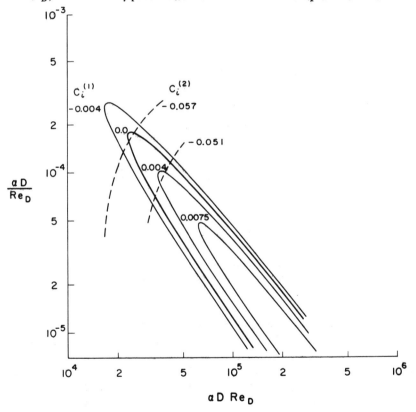

Figure 7.2. Eigenvalues as Computed:
$\alpha D/\mathrm{Re}_D$ versus $\alpha D \mathrm{Re}_D$ with c_i Parameters.

for the first mode are shown with solid lines and two curves on the "amplification surface" for the second mode are shown with dashed lines. The second-mode curve, which crosses near the minimum-critical value of $\alpha D \operatorname{Re}_D$, in Fig. 7.3 is for a c_i of -0.057, which is strong damping. A log scale has been used for $\alpha D \operatorname{Re}_D$ to compress the large range of the solutions. Since the curves approach straight lines at large values of the abscissa in both figures, a power-law dependence of form

$$c_r \propto (\alpha D \operatorname{Re}_D)^{a_1}, \qquad \alpha D / \operatorname{Re}_D \propto (\alpha D \operatorname{Re}_D)^{a_2}$$

is indicated (for large $\alpha D \operatorname{Re}_D$).

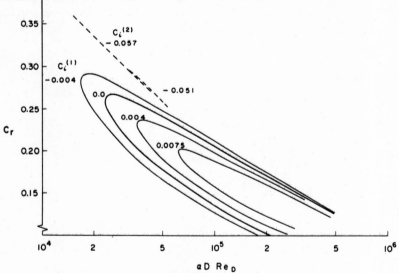

Figure 7.3. Eigenvalues as Computed:
c_r versus $\alpha D \operatorname{Re}_D$ with c_i Parameters.

The eigenvalues are plotted in more conventional form in Figs. 7.4, 7.5, and 7.6. Figure 7.4 shows αD as a function of Re_D with c_i parameters for both the first and second mode. In this and the following two figures, Re_D is on a log scale because of the large range, which for the neutral curve is from 11,560 (minimum critical) to 200,000. Amplification factors c_i of -0.004, 0, 0.004, and 0.0075 are shown. The last value was chosen because it would correspond to a contour that closes below $\operatorname{Re}_D = 200,000$ on Shen's plots [3]. This contour clearly does not close or begin to close by $Re_D = 200,000$ on the plots of this chapter. Second-mode curves for c_i of -0.057 and -0.051 are also shown in Figs. 7.4 through 7.6. The large radius of curvature of these second mode curves indicates that the neutral curve must lie at very large values of Re_D.

Figures 7.5 and 7.6 show $\beta_r \nu / U_m^2$ and c_r plotted for the same independent variable and c_i curves as in Fig. 7.4. The straightness of the c_i curves in Fig. 7.5 again indicates a power-law dependence analogous to those in

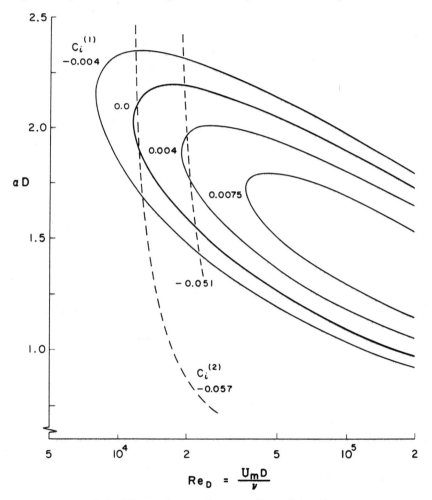

Figure 7.4.   Eigenvalues: αD versus Re_D with c_i Parameters.

Figs. 7.3 and 7.4. The separation of the first- and second-mode amplification surfaces in the c_r "direction" in parameter space for the same values of wave number αD and Reynolds number Re_D can be seen clearly in Fig. 7.6.

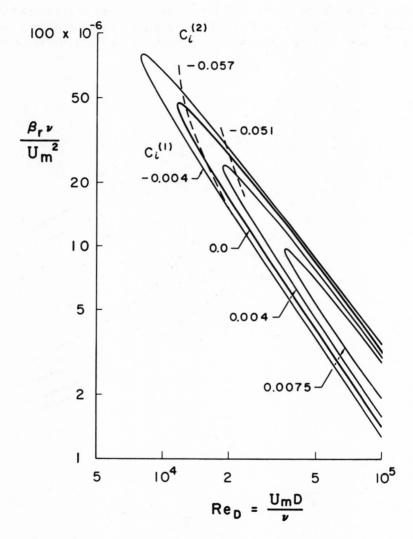

Figure 7.5.   Eigenvalues: $\beta_r \nu / U_m^2$ versus Re_D with c_i Parameters.

The last of the figures on eigenvalues, Fig. 7.7, shows the "amplification surface" for the first mode in "parameter space" (Re_D, αD, $\beta_r \nu / U_m^2$ axes). The surface is composed of curves of constant c_i, with traces of planes with values of each of the three coordinates constant. The surface has been depicted as if it were supported by a solid to make it easier to visualize in the oblique projection.

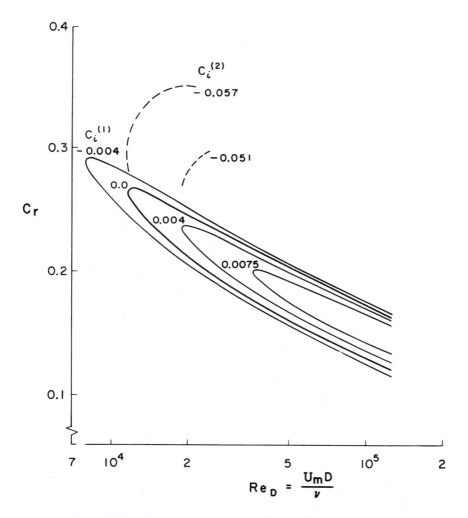

Figure 7.6.   Eigenvalues: c_r versus Re_D with c_i Parameters.

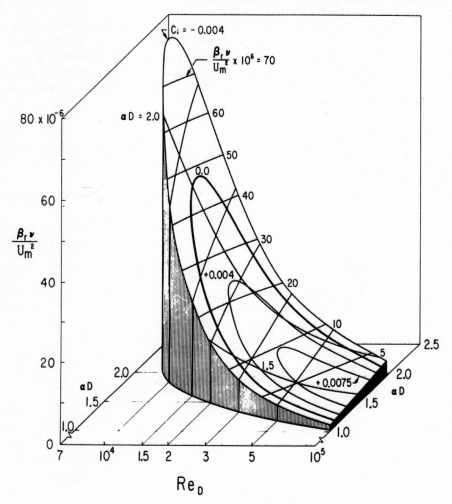

Figure 7.7 Amplification Surface.

7.8 DISTURBED FLOW FIELD PLOTS

Plots of the flow perturbation as well as several aspects of the perturbed flow field in the channel are given below grouped by case (i.e., αD, Re_D mode) rather than by type of plot. For each case, the stream function and velocity vector plots are shown for the perturbation, and for both a fixed reference frame and a reference frame moving with the propagation velocity at 20% disturbance. In addition the instantaneous local rate of energy transfer from the basic flow to the disturbance is given.

Four cases are given, one each on the first, second, third, and fourth modes. The first-mode case is near the minimum-critical Reynolds number on the neutral curve at $\alpha D = 2.02$ and $Re_D = 11,560$. For the second mode, the case is for $\alpha D = 2.0$, and $Re_D = 12,000$, with damping $c_i = -0.057$. The third mode is for $\alpha D = 2.0$, $Re_D = 12,000$, $c_i = -0.076$; and the fourth mode, for $\alpha D = 1.9$, $Re_D = 20,000$, $c_i = -1.40$. Plots of 5% disturbance are included for the first-mode minimum-critical case to indicate, for comparison, the essentially undisturbed parabolic Poiseuille profile.

All plots show the instantaneous condition at all locations in a reference frame as seen in a flash picture. The authors believe the "snapshot" approach to be more desirable than that of plotting the values that would be observed passing a given point with the stream as a function of time.

All plots have distorted coordinates, so that the x and y distance scales are different. Wave length on x and channel width on y have been taken equal, for convenience in plotting. For a typical wave number of 2, the plot should actually be π times as long as it is high.

Stream function plots are normalized so that the increase of the stream function for the undisturbed flow across the channel is from -1 to $+1$. That is, the undisturbed flow rate is normalized to 2. The percent disturbance is calculated on maximum x velocity of the disturbance and undisturbed flow center line velocity.

In the vector velocity plots, the x and y components are in the proper proportion even though the grid of points at which they are plotted has been compressed, as described above. The magnitude of the vectors has been scaled for ease of reading on each plot so that no vector will reach into a neighboring "box." However, this scaling prevents comparison of vector magnitudes between plots.

The energy transfer plot presents the quantity $-u_p v_p \, dU/dy$, which is shown in Chapter 6 to be the energy transferred per unit volume per unit time from the basic flow to the disturbance in a reference frame moving with the basic flow. The integral of this quantity over a wave length of channel must be positive for a neutrally stable disturbance, in order to make up for the additional viscous dissipation due to the disturbance. For the heavily damped higher modes, the field will be almost entirely negative, indicating that the disturbance would lose energy directly to the basic flow.

7.9 FIRST MODE MINIMUM-CRITICAL

The perturbation stream function plot for the first mode near the minimum-critical Reynolds number ($\alpha D = 2.024$, $\mathrm{Re}_D = 11{,}560$) (Fig. 7.8) shows vortices of alternate rotation filling the channel. At 5% disturbance (Fig. 7.9) a slight wave is introduced into the stream lines. At 20% disturbance (Fig. 7.10), separated regions have developed on the walls, which

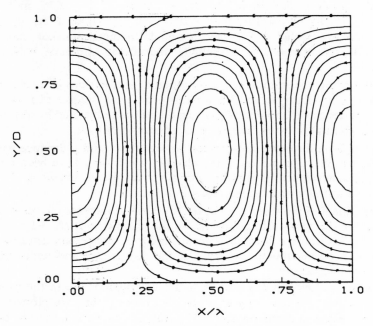

Figure 7.8.   First Mode Minimum-Critical Perturbation Stream Function.

alternate from side to side, and the stream lines in the center of the channel show a pronounced wave. In the reference frame moving with the wave velocity, c_r, the plot for 20% disturbance (Fig. 7.11), shows the separated regions apparently lifted from the wall to become alternating vortices in the flow. Vector velocity plots for the perturbation (Fig. 7.12) the 5% disturbance (Fig. 7.13), the 20% disturbance (Fig. 7.14), and the 20% disturbance in a reference frame moving with velocity c_r (Fig. 7.15) show the same separated regions in the flow as do the stream line plots.

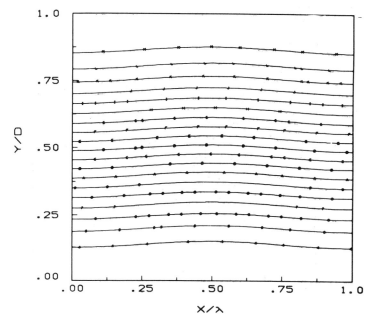

Figure 7.9. First Mode Minimum-Critical Stream
Function for 5% Disturbance.

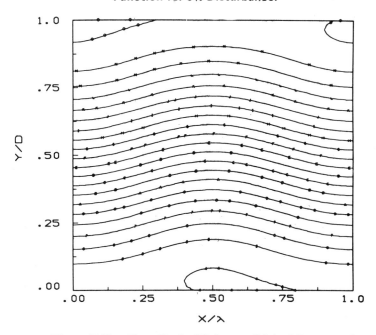

Figure 7.10. First Mode Minimum-Critical Stream
Function for 20% Disturbance.

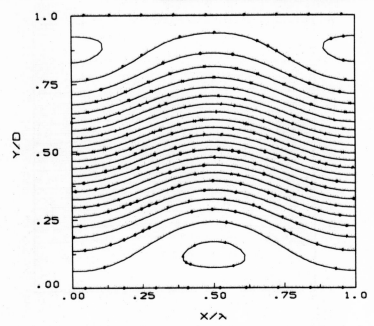

**Figure 7.11.   First Mode Minimum-Critical Stream
Function in Moving Reference Frame for 20% Disturbance.**

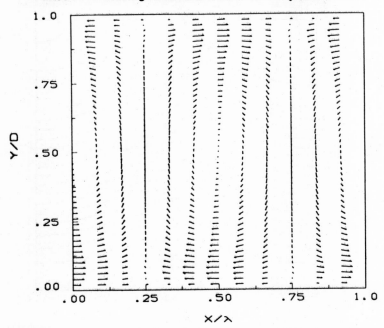

Figure 7.12.   First Mode Minimum-Critical Vector Perturbation Velocity.

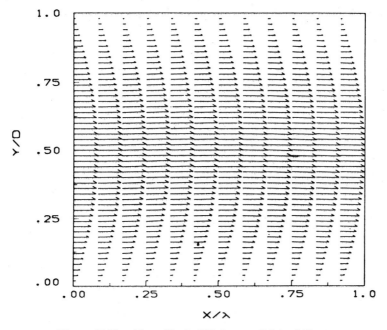

Figure 7.13.   First Mode Minimum-Critical Vector
Velocity for 5% Disturbance.

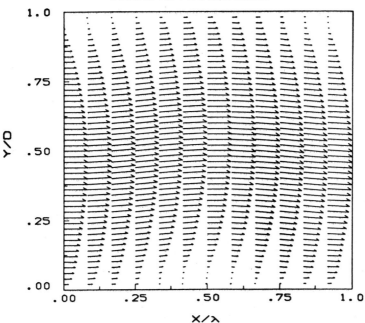

Figure 7.14.   First Mode Minimum-Critical Vector
Velocity for 20% Disturbance.

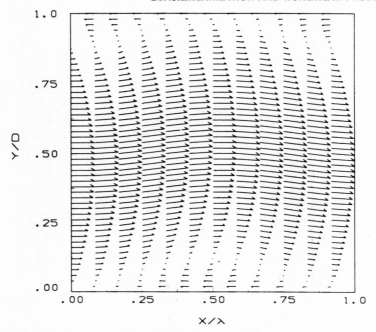

Figure 7.15. First Mode Minimum-Critical Vector
Velocity for 20% Disturbance in Moving Reference Frame.

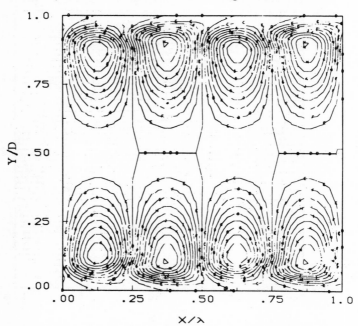

Figure 7.16. First Mode Minimum-Critical Energy Transfer.

A plot of energy transferred to the disturbance by the basic flow is shown in Fig. 7.16. The sign of the energy transfer alternates every quarter wave length and is the same on both sides of the channel. The second and fourth quarters are positive and bring in slightly more power than is lost in the first and third to account for the incremental dissipation. Magnitude of energy transfer is maximum at $\frac{1}{8}$ channel width from the walls on each side, dropping rapidly to zero at the walls and slowly to zero at the center line. The higher modes will be found to have increasingly smaller regions, where the magnitude of energy transfer is large, due to the steeper slopes in the eigenfunctions.

7.10 SECOND MODE

A point at $\alpha D = 1.952$ and $\mathrm{Re}_D = 12,000$ on the $c_i = -0.057$ curve shown in Fig. 7.4 was chosen to represent the second mode because this point is close to the minimum Reynolds number on the first-mode neutral curve. Since all of the other cases calculated were also in the heavily damped region, the velocity and stream functions would be very similar.

The perturbation stream function shown in Fig. 7.17 shows the channel divided into four cells per wave length. Although there is a straight dividing line down the center of the channel, the transverse dividing lines are not straight but show salients in the upstream direction close to the walls. This feature is attributed to the imaginary part of the eigenfunction, which is large near the wall. Figures 7.18 and 7.19 show the stream functions for the mean flow plus 20% disturbance in both the fixed and moving reference frames, with two separated regions opposite each other and a pronounced bulge in the rest of the stream lines.

Concentration of the disturbance near the walls is even more apparent in the perturbation velocity plot (Fig. 7.20) than in the perturbation stream function. The flow near the walls is compensated by a return flow with a smaller velocity that is nearly constant across most of the channel. Velocity plots for 20% disturbance, for both fixed and moving reference frames, are also shown in Figs. 7.21 and 7.22.

Although this is a heavily damped case, the four principal regions of positive energy transfer shown in Fig. 7.23 are nearly as large as the negative regions. Most of the energy transfer, both positive and negative, lies in two strips within slightly more than $\frac{1}{8}$ channel width of either wall. The energy transfer in the remaining center portion is negligible. Positive and negative regions are no longer confined to quarter wave lengths, as in the first mode, because of the large relative importance of the imaginary part of the eigenfunction.

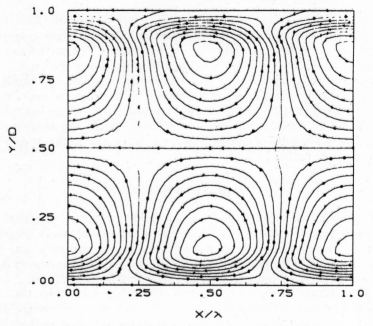

Figure 7.17.   Second Mode $\alpha D = 1.952$, $Re_D = 12{,}000$,
$c_i = -0.057$, **Perturbation Stream Function.**

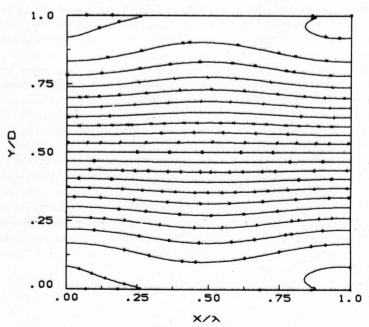

Figure 7.18.   Second Mode $\alpha D = 1.952$, $Re_D = 12{,}000$,
$c_i = -0.057$; **Stream Function for 20% Disturbance.**

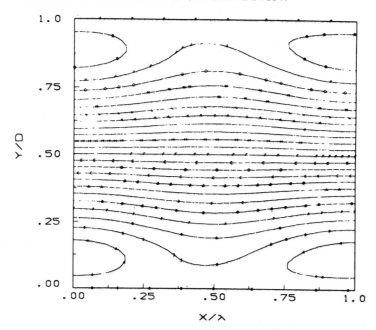

Figure 7.19. Second Mode $\alpha D = 1.952$, $\text{Re}_D = 12{,}000$,
$c_i = -0.057$; Stream Function for 20%,
Disturbance in Moving Reference Frame.

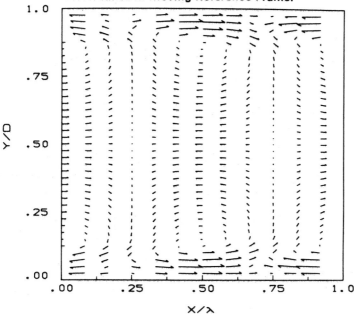

Figure 7.20. Second Mode $\alpha D = 1.952$, $\text{Re}_D = 12{,}000$,
$c_i = -0.057$; Vector Perturbation Velocity.

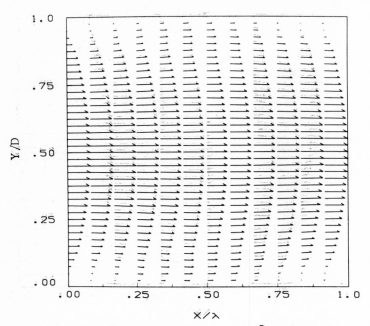

Figure 7.21. Second Mode $\alpha D = 1.952$, $\text{Re}_D = 12,000$, $c_i = -0.057$; Vector Velocity for 20% Disturbance.

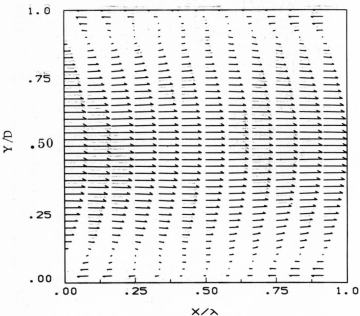

Figure 7.22. Second Mode $\alpha D = 1.952$, $\text{Re}_D = 12,000$, $c_i = -0.057$; Vector Velocity for 20% Disturbance in Moving Reference Frame.

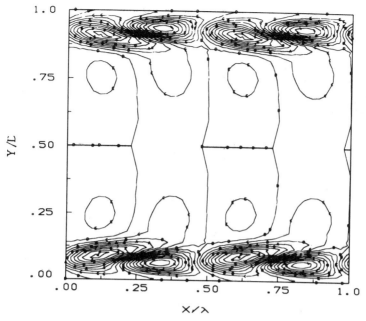

Figure 7.23. Second Mode $\alpha D = 1.952$, $Re_D = 12,000$,
$c_i = -0.057$; Energy Transfer.

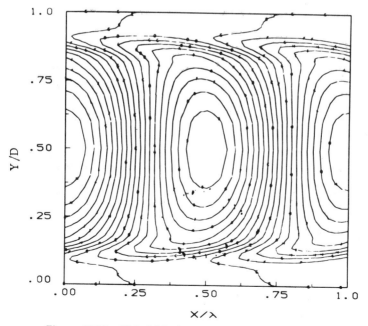

Figure 7.24. Third Mode $\alpha D = 2.000$, $Re_D = 12,000$,
$c_i = -0.0763$; Perturbation Stream Function.

7.11   THIRD MODE

The third mode has the same symmetry as the first mode. As a result there is a strong resemblance between the perturbation stream function for the third mode (Fig. 7.24) and that for the first mode (Fig. 7.8). However, the third-mode stream function has salients in the outer contours that point in an upstream direction near the walls rather than the nearly elliptical contour of the first mode. Because of the higher propagation velocity, the perturbation velocity is small for the approximately 0.1 channel width closest to each wall. This in turn makes the separated regions for 20% disturbance in the fixed reference frame (Fig. 7.25) small. The vortices in the moving reference frame (Fig. 7.26) are also displaced toward the center line. The streamlines in this last figure are also somewhat distorted on the edge of the vortex regions.

Regions of high perturbation velocity of about $\frac{1}{8}$ channel width from the walls are seen in Fig. 7.27. These regions overlap at $\frac{1}{4}$ and $\frac{3}{4}$ wave length stations. The high perturbation velocity regions result in pronounced bumps in the fixed reference frame total velocity plot (Fig. 7.28).

The energy transfer plot (Fig. 7.29) for the third mode shows regions or large loss to the mean flow corresponding to the regions of large perturbation velocity. Positive energy transfer regions have almost disappeared. This mode and the fourth mode are unlikely on an energy basis, but they may contribute to nonlinear effects occurring before the onset of turbulence, as suggested by Eckhaus [4].

7.12   FOURTH MODE

The fourth mode, which has the same symmetry as the second mode, results in an even more distorted flow than the third mode, as would be expected. Plots corresponding to those shown for the lower modes are shown in Figs. 7.30 through 7.35. High-velocity bands are again to be noted near the walls.

One is led to infer that these bands of high velocity in the higher modes may be related to the tongues that have been experimentally observed by Betchov and Criminale, Jr. [5] in the development of waves in laminar boundary layers.

Additional plots may be found in the report by Radbill and van Driest [6], on which this chapter is based.

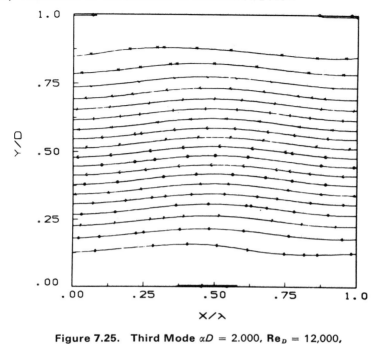

Figure 7.25. Third Mode $\alpha D = 2.000$, $Re_D = 12,000$, $c_i = -0.0763$; **Stream Function for 20% Disturbance.**

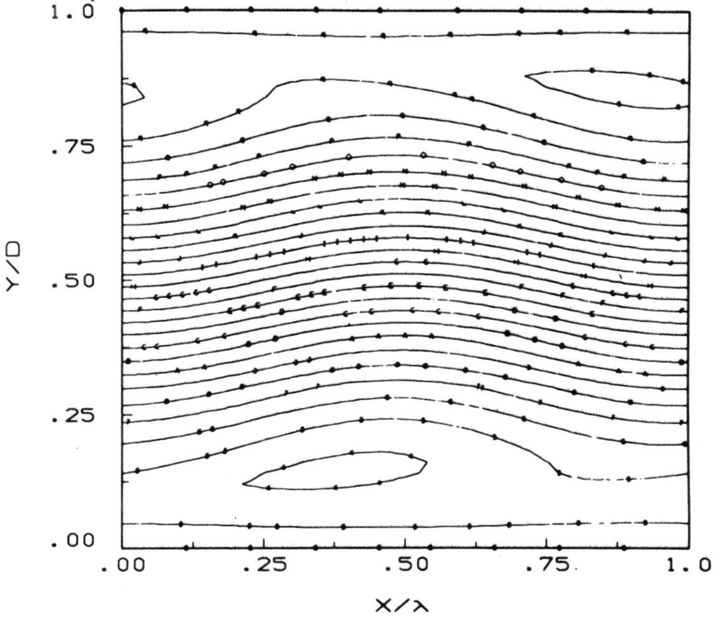

Figure 7.26. Third Mode $\alpha D = 2.00$, $Re_D = 12,000$, $c_i = -0.0763$; **Stream Function for 20% Disturbance in Moving Reference Frame.**

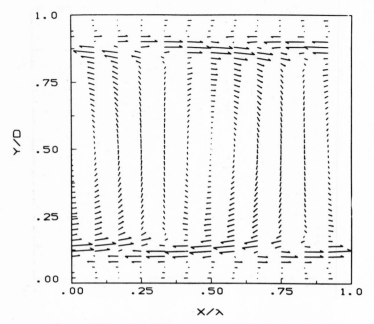

Figure 7.27. Third Mode $\alpha D = 2.000$, $\mathbf{Re}_D = 12{,}000$,
$c_i = -0.0763$; Vector Perturbation Velocity.

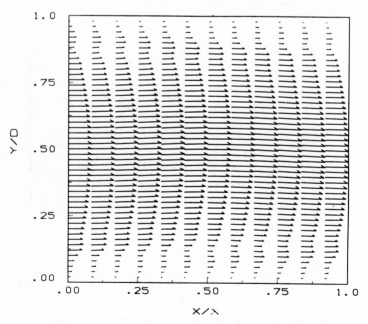

Figure 7.28. Third Mode $\alpha D = 2.000$, $\mathbf{Re}_D = 12{,}000$,
$c_i = 0.0763$; Vector Velocity for 20% Disturbance.

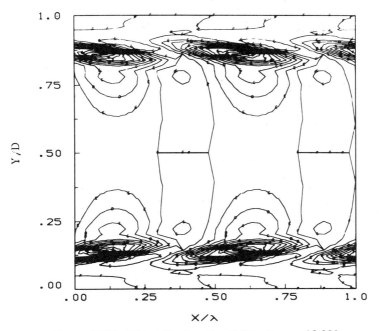

Figure 7.29. **Third Mode** $\alpha D = 2.000$, **Re**$_D = 12,000$,
$c_i = -0.0763$; **Energy Transfer.**

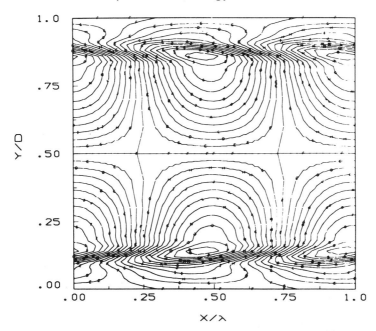

Figure 7.30. **Fourth Mode** $\alpha D = 1.900$, **Re**$_D = 20,000$,
$c_i = -0.140$; **Perturbation Stream Function.**

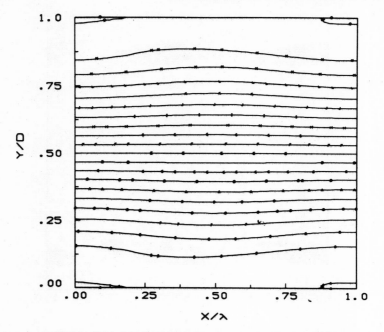

Figure 7.31. Fourth Mode $\alpha D = 1.900$, $Re_D = 20,000$,
$c_i = -0.140$; Stream Function for 20% Disturbance.

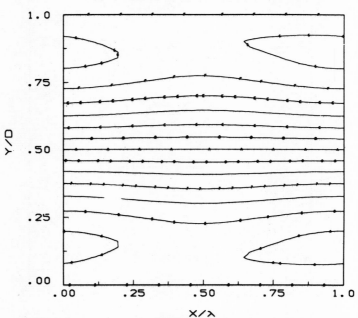

Figure 7.32. Fourth Mode $\alpha D = 1.900$, $Re_D = 20,000$,
$c_i = -0.140$; Stream Function for 20%
Disturbance in Moving Reference Frame.

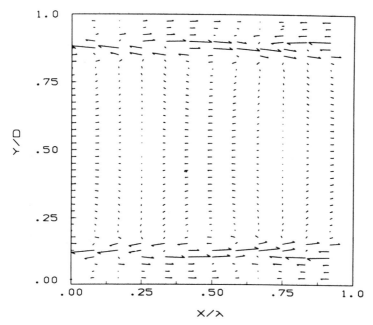

Figure 7.33. Fourth Mode $\alpha D = 1.900$, **Re**$_D$ = 20,000,
$c_i = -0.140$; **Vector Perturbation Velocity.**

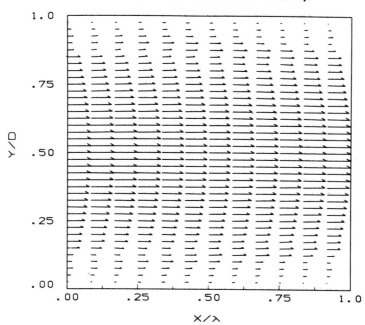

Figure 7.34. Fourth Mode $\alpha D = 1.900$, **Re**$_D$ = 20,000,
$c_i = -0.140$; **Vector Velocity for 20% Disturbance.**

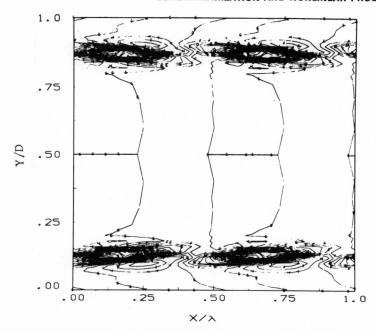

Figure 7.35.   Fourth Mode $\alpha D = 1.900$, **Re**$_D$ $= 20{,}000$,
$c_i = -0.140$; **Energy Transfer.**

7.13 CONCLUSION

Transformation of the differential equations by use of a new "stretched" independent variable has been found in this chapter to be an effective means of reducing the growth of unwanted solutions that cause the homogeneous solutions to become dependent. By removing the explicit dependence on the large parameter ($\alpha \, \mathrm{Re}_D$) and making all of the terms in the equation the same order of magnitude, the amplification of the unwanted components of the solution is nearly eliminated. The use of transformations to reduce the instability toward parasitic solutions is an important supplement to the orthogonalization introduced in earlier chapters.

Convergence of the quasilinearization algorithm to a particular member of a set of multiple solutions of an eigenvalue problem was found to be controlled by the initial approximation in parameter space. The selection of a desired member or mode from the set of solutions was facilitated by restricting the changes in the approximations to the eigenvalues during the early iterations, and by exploiting symmetry or asymmetry of the desired branch or mode of the solution. The ability to obtain a particular branch of a multiple valued solution is important because such solutions are very common in various branches of mechanics.

REFERENCES

1. P. R. Nachtsheim, "An Initial Value Method for the Numerical Treatment of the Orr-Sommerfeld Equation for the Case of Plane Poiseuille Flow," NASA TN D-2414 (1964).
2. W. H. Reid, "The Stability of Parallel Flows in Basic Developments in Fluid Dynamics," Vol. 1, pp. 260–263. Academic Press, New York, 1965.
3. S. F. Shen, Calculated amplified oscillations in plane Poiseuille and Blasius flows. *J. Aeron. Sci.* **21**, 62–64 (1954).
4. W. Eckhaus, "Studies in Non-Linear Stability Theory." Springer-Verlag, Berlin and Heidelberg, 1965.
5. R. Betchov and W. O. Criminale, Jr., "Stability of Parallel Flows," pp. 133–134. Academic Press, New York, 1967.
6. J. R. Radbill and E. R. van Driest, "Stability of Two-Dimensional Poiseuille Flow," AFOSR 67-2420 (1967).

Chapter 8

OPTIMUM ORBITAL TRANSFER
WITH "BANG-BANG" CONTROL

8.1 INTRODUCTION

As a final illustration of the power of quasilinearization, this chapter considers a particularly sensitive optimum control problem. This example points out the extreme care that is sometimes necessary to achieve satisfactory numerical results. It also considers several "tricks" that were required to obtain adequate convergence.

Certain realistic treatments of the optimum orbital transfer problem lead to a variational formulation wherein the differential equations have no exact closed-form solution. Prior experience with the particular two point boundary-value problem considered here (see Jurovics [1]) indicated that it could not be solved by conventional numerical search methods. Reformulation and application of quasilinearization (see Kalaba [2], Bellman et al. [3], and McGill and Kenneth [4]) allowed the successful computation of optimal "finite-thrust" transfer trajectories between arbitrary pairs of coplanar elliptical orbits. However, successful application of quasilinearization was found to depend upon the proper use of initial conditions derived from an optimum two-impulse transfer maneuver (see McCue [5] and [6], McCue and Bender [7] and [8].)

8.2 OPTIMIZATION PROBLEM

The problem to be considered here involves transfers between a pair of coplanar orbits defined by their semi-latera recta p_1, p_2, eccentricities e_1, e_2, and arguments of perigee ω_1, ω_2. What is required is the determination of that trajectory which results in an orbital transfer with minimum fuel expenditure. The formulation is a modified version of a three-dimensional derivation originated by Jurovics [1] and is similar to that presented by Leitmann [9]. Solution of this optimization problem involves the minimization of a functional that is a function of only the boundary values of the state variables, i.e., position, velocity, mass, and time.

The function to be minimized is the characteristic velocity

$$G = c \ln [m(0)/m(T)] \tag{8.1}$$

where

$$G = G(X_i(0), X_i(T)) \tag{8.2}$$

and, where the state variables are

$$X_i = \{r, \phi, r', \phi', m\} \tag{8.3}$$

In the above expressions c is the effective exhaust velocity, r the radius, ϕ the central angle, m the mass, and the times 0 and T refer to initial and final points of the trajectory.

The rocket and its environment are defined in accordance with the following assumptions: (1) The rocket is a variable mass particle; (2) the thrust magnitude (F) is a linear function of the mass flow rate (β):

$$F = c\beta = -cm' \tag{8.4}$$

(3) the vehicle is capable of thrust direction and throttle control, and the control is instantaneous; (4) further, the transfer maneuver is between two orbits about a single planet with a spherically symmetrical central gravitational field.

EQUATIONS OF MOTION. In polar coordinates (Fig. 8.1), the two second-order equations of motion are

$$r'' - r\phi'^2 + \mu/r^2 = (F/m) \cos v \tag{8.5}$$
$$r\phi'' + 2r'\phi' = (F/m) \sin v \tag{8.6}$$

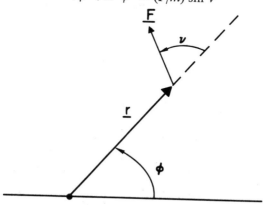

Figure 8.1. Coordinate System Definition.

where μ is the gravitational constant and v the steering angle measured from the local vertical.

These equations may be reduced to first-order form, where the new variables ρ and y are defined as

$$\rho = r' \tag{8.7}$$

$$y = \phi' \tag{8.8}$$

$$\rho' = ry^2 - \mu/r^2 + (c\beta/m)\cos v \tag{8.9}$$

$$y' = -(2\rho y/r) + (c\beta/mr)\sin v \tag{8.10}$$

$$m' = -\beta \tag{8.11}$$

EULER-LAGRANGE EQUATIONS. The optimum path must satisfy the equations of motion (8.7) to (8.11). In addition, for most conventional rockets, the solution is subject to the following constraints:

$$c = \text{constant} \tag{8.12}$$

$$\beta_{\min} \leqslant \beta \leqslant \beta_{\max} \tag{8.13}$$

where $\beta_{\min} = 0$, and $\beta_{\max}$ is specified.

In the problem considered here one utilizes the following additional constraint to impose "bang-bang" control of mass flow rate:

$$\beta(\beta - \beta_{\max}) = 0 \tag{8.14}$$

If G is to possess an extremum subject to the constraints imposed by Eqs. (8.7) to (8.12) and (8.14), one must require the first variation of the constrained functional to vanish. The following Euler-Lagrange differential equations for the Lagrange multipliers result:

$$\lambda_1' = -\lambda_4\left(y^2 + \frac{2\mu}{r^3}\right) - \lambda_6\left(\frac{2\rho y}{r^2} - \frac{c\beta\sin v}{mr^2}\right) \tag{8.15}$$

$$\lambda_3' = 0 \tag{8.16}$$

$$\lambda_4' = -\lambda_1 + 2\lambda_6 y/r \tag{8.17}$$

$$\lambda_6' = -\lambda_3 - 2\lambda_4 ry + 2\lambda_6 \rho/r \tag{8.18}$$

$$\lambda_7' = \frac{c\beta}{m^2}\left[\lambda_4\cos v + \frac{\lambda_6\sin v}{r}\right] \tag{8.19}$$

The differential equation for λ_7 may also be written in the useful form

$$\lambda_7' = (\beta/m)[k+\lambda_7] \tag{8.20}$$

The "switching function" k, which appears in Eq. (8.20), governs thrust on-off control and is defined as

$$k = \frac{c}{m}\left(\lambda_4 \cos v + \frac{\lambda_6 \sin v}{r}\right) - \lambda_7 \tag{8.21}$$

where

$$k > 0 \Rightarrow \beta = \beta_{max} \tag{8.22}$$

$$k < 0 \Rightarrow \beta = 0 \tag{8.23}$$

Equations for the steering angle v are as follows:

$$\sin v = \lambda_6/D \tag{8.24}$$

$$\cos v = \lambda_4 r/D \tag{8.25}$$

where

$$D = (\lambda_6^2 + \lambda_4^2 r^2)^{1/2} \tag{8.26}$$

Clearly, the steering angle has no physical significance when the vehicle is on a coasting arc.

Since the differential equations do not involve time explicitly, one obtains a first integral (Hamiltonian):

$$\lambda_1 r' + \lambda_3 \phi' + \lambda_4 \rho' + \lambda_6 y' + \lambda_7 m' = A \tag{8.27}$$

where

$$A = \text{constant} \tag{8.28}$$

These last expressions may be used to replace one of the Euler-Lagrange differential equations and thereby reduce the order of the system by one.

BOUNDARY CONDITIONS. Note that the system is described by ten equations for the variables

$$r, \phi, \rho, y, m, \lambda_1, \lambda_3, \lambda_4, \lambda_6, \lambda_7$$

This system thus requires ten boundary conditions. The seven specified by the physics of the problem are

$$p_1, e_1, \omega_1, p_2, e_2, \omega_2, m(0)$$

The remaining boundary conditions can be derived from the transversality contition:[†]

$$dG + [\lambda_1 \, dr + \lambda_3 \, d\phi + \lambda_4 \, d\rho + \lambda_6 \, dy + \lambda_7 \, dm - A \, dt]_0^T = 0 \qquad (8.29)$$

One may then obtain the following additional boundary conditions (see Jurovics [1]):

$$\lambda_7 = \frac{c}{m}, \qquad t = T \qquad (8.30)$$

$$[\lambda_1 \, dr + \lambda_3 \, d\phi + \lambda_4 \, d\rho + \lambda_6 \, dy]_0^T = 0 \qquad (8.31)$$

A somewhat different form of Eq. (8.27) may also be derived (see Jurovics [1]):

$$[r'\lambda_1 + \phi'\lambda_3 + \rho'\lambda_4 + y'\lambda_6 + m'\lambda_7]_0^T = \beta k \qquad (8.32)$$

From this form of the equation, it is clear that the Hamiltonian [A in Eq. (8.20)] must be equal to βk at the end points. Further, if $A \neq 0$ at $t = 0$, then Eq. (8.24) implies that $k(0) = k(T)$.

Having obtained a system of ten first-order ordinary differential equations, which must yield the required optimal trajectory over a specified time interval, one next observes that the problem is of mixed end-value nature and only the five values of the state variables are known at the initial point (or final point). However, it is well known that the λ_i can be scaled by a positive constant. (See Leitmann [9].) Thus, by assigning an appropriate initial value to λ_1, the number of unknown initial conditions was reduced to four.

CORNER CONDITIONS. For this particular problem, the corner conditions are such that the multipliers associated with each of the state variables and A, the first integral, must have the same value immediately preceding and following a corner:

$$(\lambda_i) - \; = (\lambda_i) + \qquad (8.33)$$

$$A - \; = A + \qquad (8.34)$$

8.3 QUASILINEARIZATION

Having obtained a nonlinear two point boundary-value problem, the method of quasilinearization (see Kalaba [2]) may be used to generate the required numerical solutions. The previously derived differential equations may be written as a set of ten first-order equations, each of which may be

[†] The notation [ ]$_0^T$ indicates that the equation applies at the ends of the interval.

considered to be one component of the vector equation:

$$\mathbf{X}' = \mathbf{g}(\mathbf{X}) \tag{8.35}$$

where

$$
\mathbf{X} =
\begin{bmatrix}
r \\
\phi \\
\rho \\
y \\
m \\
\lambda_1 \\
k \\
\lambda_4 \\
\lambda_6 \\
\lambda_3
\end{bmatrix}
\qquad
\mathbf{g} =
\begin{bmatrix}
\rho \\
y \\
ry^2 - \dfrac{\mu}{r^2} + \dfrac{c\beta}{m}\cos v \\
-\dfrac{2\rho y}{r} + \dfrac{c\beta}{mr}\sin v \\
-\beta \\
-\lambda_4\left(y^2 + \dfrac{2\mu}{r^3}\right) - \lambda_6\left(\dfrac{2\rho y}{r^2} - \dfrac{c\beta}{mr^2}\sin v\right) \\
-\dfrac{c}{mD}\left(r\lambda_1\lambda_4 + \dfrac{\lambda_3\lambda_6}{r} - \dfrac{\rho\lambda_6^2}{r^2}\right) \\
-\lambda_1 + \dfrac{2\lambda_6 y}{r} \\
-\lambda_3 - 2\lambda_4 ry + \dfrac{2\lambda_6\rho}{r} \\
0
\end{bmatrix}
\tag{8.36}
$$

Note that, in the above equations, an expression for the switching function k has been derived from Eq. (8.20) and substituted for λ_7. This substitution was useful because it replaced an unknown function with one with properties partially defined by the optimum two-impulse transfer maneuver. Since λ_3 is constant throughout a given trajectory, it may be regarded as a parameter whose time history may be obtained without resorting to numerical integration. In order to achieve computational economies with the authors' general quasilinearization program, it was required that λ_3 be assigned to the last position in the vectors of Eq. (8.36).

Having established the differential equation, we now write the fundamental equation of quasilinearization (see Kalaba [2] and Bellman *et al.* [3]):

$$X_i^{\prime(n+1)}(t) = g_i(\mathbf{X}^{(n)}(t)) + \sum_{j=1}^{N} [X_j^{(n+1)}(t) - X_j^{(n)}(t)] \frac{\partial g_i(\mathbf{X}^{(n)}(t))}{\partial X_j}$$

$$(i = 1, 2, \ldots, N) \qquad (8.37)$$

where N is the number of differential equations. In this case the Jacobian matrix of partial derivatives is rather involved [Eq. (8.38) see p. 151]. The more lengthy partial derivatives appearing in the Jacobian matrix appear as Eqs. (8. 39).

$$\frac{\partial \rho'}{\partial r} = y^2 + \frac{2\mu}{r^3} + \frac{c\beta\lambda_4\lambda_6^2}{mD^3}$$

$$\frac{\partial y'}{\partial r} = \frac{2\rho y}{r^2} - \frac{c\beta\lambda_6}{mr^2 D}\left(1 + \frac{r^2\lambda_4^2}{D^2}\right)$$

$$\frac{\partial \lambda_1'}{\partial r} = \frac{6\mu\lambda_4}{r^4} + \frac{4\rho y\lambda_6}{r^3} - \frac{c\beta\lambda_6^2}{mr^3 D}\left(2 + \frac{r^2\lambda_4^2}{D^2}\right)$$

$$\frac{\partial k'}{\partial r} = \frac{c}{mD}\left[\frac{\lambda_3\lambda_6}{r^2} - \frac{2\rho\lambda_6^2}{r^3} - \frac{1}{D^2}\left(\lambda_1\lambda_4\lambda_6^2 - \lambda_3\lambda_4^2\lambda_6 + \frac{\rho\lambda_4^2\lambda_6^2}{r}\right)\right]$$

$$\frac{\partial \lambda_1'}{\partial y} = 2y\lambda_4 - \frac{2\rho\lambda_6}{r^2}$$

$$\frac{\partial k'}{\partial m} = \frac{c}{m^2 D}\left(r\lambda_1\lambda_4 + \frac{\lambda_3\lambda_6}{r} - \frac{\rho\lambda_6^2}{r^2}\right) \qquad (8.39)$$

$$\frac{\partial \lambda_1'}{\partial \lambda_4} = -y^2 - 2\frac{\mu}{r^3} - \frac{c\beta\lambda_6^2\lambda_4}{mD^3}$$

$$\frac{\partial k'}{\partial \lambda_4} = \frac{c\lambda_6}{mD^3}(-r\lambda_1\lambda_6 + r\lambda_3\lambda_4 - \rho\lambda_4\lambda_6)$$

$$\frac{\partial \lambda_1'}{\partial \lambda_6} = \frac{-2\rho y}{r^2} + \frac{c\beta\lambda_6}{mr^2 D}\left(2 - \frac{\lambda_6^2}{D^2}\right)$$

$$\frac{\partial k'}{\partial \lambda_6} = \frac{c}{mD^3}\left(-r\lambda_3\lambda_4^2 + \frac{\rho\lambda_6^3}{r^2} + 2\rho\lambda_6\lambda_4^2 + r\lambda_6\lambda_1\lambda_4\right)$$

$$
\left[\frac{\partial g_i}{\partial X_j}\right] =
$$

$$
\begin{bmatrix}
0 & 0 & 0 & 0 & 0 & 0 & -\dfrac{c\lambda_6}{rmD} & 0 & -1 & 0 \\[2ex]
0 & 0 & -\dfrac{c\beta r\lambda_4\lambda_6}{mD^3} & \dfrac{c\beta r\lambda_4^2}{mD^3} & 0 & \dfrac{\partial\lambda_1'}{\partial\lambda_6} & \dfrac{\partial k'}{\partial\lambda_6} & \dfrac{2y}{r} & \dfrac{2\rho}{r} & 0 \\[2ex]
0 & 0 & \dfrac{c\beta r\lambda_6^2}{mD^3} & -\dfrac{c\beta r\lambda_4\lambda_6}{mD^3} & 0 & \dfrac{\partial\lambda_1'}{\partial\lambda_4} & \dfrac{\partial k'}{\partial\lambda_4} & 0 & -2ry & 0 \\[2ex]
0 & 0 & 0 & 0 & 0 & 0 & 0 & 0 & 0 & 0 \\[2ex]
0 & 0 & 0 & 0 & 0 & 0 & -\dfrac{c\lambda_4 r}{mD} & -1 & 0 & 0 \\[2ex]
0 & 0 & -\dfrac{c\beta\lambda_4 r}{m^2 D} & -\dfrac{c\beta\lambda_6}{m^2 r D} & 0 & -\dfrac{c\beta\lambda_6^2}{m^2 r^2 D} & \dfrac{\partial k'}{\partial m} & 0 & 0 & 0 \\[2ex]
0 & 1 & 2ry & -\dfrac{2\rho}{r} & 0 & \dfrac{\partial\lambda_1'}{\partial y} & 0 & \dfrac{2\lambda_6}{r} & -2\lambda_4 r & 0 \\[2ex]
1 & 0 & 0 & -\dfrac{2y}{r} & 0 & -\dfrac{2y\lambda_6}{r^2} & \dfrac{c\lambda_6^2}{mD r^2} & 0 & \dfrac{2\lambda_6}{r} & 0 \\[2ex]
0 & 0 & 0 & 0 & 0 & 0 & 0 & 0 & 0 & 0 \\[2ex]
0 & 0 & \dfrac{\partial\rho'}{\partial r} & \dfrac{\partial y'}{\partial r} & 0 & \dfrac{\partial\lambda_1'}{\partial r} & \dfrac{\partial k'}{\partial r} & -\dfrac{2\lambda_6 y}{r^2} & -2\lambda_4 y-\dfrac{2\lambda_6\rho}{r^2} & 0
\end{bmatrix}
\tag{8.38}
$$

Recalling the analysis of Chapter 2 we note that because Eq. (8.37) is linear in the $X_i^{(n+1)}$, solutions may be added to satisfy all the boundary conditions. By requiring a particular solution $P_i(t)$ to satisfy the known initial boundary conditions, it is necessary to generate $P_i(t)$ plus as many homogeneous solutions $H_{ij}(t)$ as there are unknown initial boundary conditions B_i. A modification of Eq. (1.25) that gives the correct number of equations to solve for the combination coefficients C_j in the remainnig four boundary conditions is thus obtained:

$$B_i(T) - P_i(T) = \sum_{j=1}^{4} C_j H_{ij}(T) \qquad (i = 1, 2, 3, 4) \qquad (8.40)$$

As before, the $(n+1)$'st approximation, $X_i^{(n+1)}(t)$, is evaluated by summing the stored values of $P_i(t)$ and $H_{ij}(t)$:

$$X_i^{(n+1)}(t) = P_i(t) + \sum_{j=1}^{4} C_j H_{ij}(t) \qquad (i = 1, 2, \ldots, N) \qquad (8.41)$$

We have previously observed that if the process converges it will do so quadratically. However, in order for the process to converge, one must have a sufficiently accurate first approximation, $\mathbf{X}^{(0)}(t)$, to the time history of the solution vector. Convergence may be examined by evaluating the following relationships at the end of each iteration:

$$\xi_i(\mathbf{X}^{(n)}(t), \mathbf{X}^{(n+1)}(t)) = \max_t |X_i^{(n)}(t) - X_i^{(n+1)}(t)| \qquad (i = 1, 2, \ldots, N) \quad (8.42)$$

The abrupt changes in mass flow rate require special consideration. Whenever the switching function exhibits a zero ("switch point"), the mass flow rate ($m' = -\beta$) must undergo a discontinuous change. This property can be expressed by employing a unit step function in the definition of β (see Friedman [10]):

$$\beta = \beta_0 u[k^{(n)}(t)] \qquad (8.43)$$

where

$$u[k^{(n)}(t)] = \int_{k-}^{k+} \delta[k^{(n)}(t)] \, dk \qquad (8.44)$$

At a switch point ($k^{(n)}(t) = 0$ and $t = t_s$), Eq. (8.37) must be integrated as follows:

$$\int_{t_s-}^{t_s+} X_i'^{(n+1)}(t) \, dt = \int_{t_s-}^{t_s+} \left\{ g_i(\mathbf{X}^{(n)}(t)) + \sum_{j=1}^{N} [X_j^{(n+1)}(t) - X_j^{(n)}(t)] \frac{\partial g_i(\mathbf{X}^{(n)}(t))}{\partial X_j} \right\} dt$$

$$(i = 1, 2, \ldots, N) \qquad (8.45)$$

The $g_i(\mathbf{X}^{(n)}(t))$ term offers no net contribution and may therefore be excluded from further consideration. Next, observe that, at a switch point, the Jacobian matrix will have some nonzero terms appearing in the column containing partial derivatives with respect to the switching function k. Note that this is only true at the switch points and that Eq. (8.38) is valid elsewhere. Thus, Eq. (8.45) becomes

$$\Delta X_i^{(n+1)}(t_s) = \int_{t_{s-}}^{t_{s+}} [k^{(n+1)}(t) - k^{(n)}(t)] \frac{\partial g_i(\mathbf{X}^{(n)}(t))}{\partial k} dt$$

$$(i = 1, 2, \ldots, N) \qquad (8.46)$$

Since the switching time is determined by the n'th iteration wherein $k^{(n)}(t) = 0$, one obtains

$$\Delta X_i^{(n+1)}(t) = \int_{t_{s-}}^{t_{s+}} k^{(n+1)}(t) \frac{\partial g_i(\mathbf{X}^{(n)}(t))}{\partial k} dt \qquad (i = 1, 2, \ldots, N) \qquad (8.47)$$

At this point it should be noted that only those terms in Eq. (8.36) which contain mass flow rate will contribute to the solution at a discontinuity. Substitution of Eqs. (8.36), (8.43), and (8.44) into Eq. (8.47) yields expressions of the following form:

$$\Delta X_i^{(n+1)}(t) = S_i \beta_0 k^{(n+1)}(t) \int_{t_{s-}}^{t_{s+}} \frac{\partial}{\partial k} \int_{k-}^{k+} \delta[k^{(n)}(t)] \, dk \, dt \qquad (8.48)$$

or

$$\Delta X_i^{(n+1)}(t) = S_i \beta_0 k^{(n+1)}(t) \int_{t_{s-}}^{t_{s+}} \delta[k^{(n)}(t)] \, dt \qquad (i = 1, 2, \ldots, N) \qquad (8.49)$$

where the S_i represent arbitrary constants.

In order to integrate with respect to the argument of the delta function, we adopt the following definitions:

$$k = F(t), \qquad t = F^{-1}(k) \qquad (8.50)$$

$$dt = (d/dk)F^{-1}(k) \, dk$$

Equation (8.49) may now be rewritten and integrated with respect to k:

$$\Delta X_i^{(n+1)}(t) = S_i \beta_0 k^{(n+1)}(t) \int_{k-}^{k+} \delta[k^{(n)}(t)] \frac{d}{dk} F^{-1}(k) \, dk \qquad (8.51)$$

$$\Delta X_i^{(n+1)}(t) = S_i \beta_0 k^{(n+1)}(t) \frac{dt}{dk} \int_{k-}^{k+} \delta[k^{(n)}(t)] \, dk \tag{8.52}$$

$$\Delta X_i^{(n+1)}(t) = S_i \beta_0 k^{(n+1)}(t)/k'(t) \qquad (i = 1, 2, \ldots, N) \tag{8.53}$$

Equation (8.53) can now be applied to derive the following final expressions for the contributions to the solution at a corner point:

$$\Delta \rho = \Delta t (c\beta/m) \cos v \tag{8.54}$$

$$\Delta y = \Delta t (c\beta/mr) \sin v \tag{8.55}$$

$$\Delta m = -\beta \Delta t \tag{8.56}$$

$$\Delta \lambda_1 = \Delta t (c\beta \lambda_6/mr^2) \sin v \tag{8.57}$$

$$\Delta t = \frac{k^{(n+1)}(t_s)}{[k'(t_s)]} \tag{8.58}$$

During the integration of the particular and homogeneous solutions required by Eq. (8.40), $k^{(n+1)}(t_s)$ will, in general, be nonzero at the switch points determined from the n'th iteration. The quantity Δt, defined in Eq. (8.58), may be regarded as an incremental change in the length of a burn period called for since $k^{(n+1)}(t_s)$ is now nonzero at the switch points. The incremental changes defined by Eqs. (8.54) through (8.57) compensate for the changes that will result from these small changes in burning time. The need for the absolute value sign can be established by considering the physical implications of a nonzero value of $k^{(n+1)}(t)$ at points where $k^{(n)}(t) = 0$.

8.4 NUMERICAL SOLUTION BY QUASILINEARIZATION

8.4.1 Computation Techniques

Equations (8.1) through (8.58) were programmed in FORTRAN IV for solution by an IBM 7094 digital computer. A double precision integration program (Chapter 9) employing Runge-Kutta starting procedures and a variable step-size difference integration scheme was utilized. The current approximation to the solution, $X^{(n)}(t)$, as well as $P(t)$ and the $H_j(t)$, were generated by integration and stored at fixed tabular intervals. During the integration, it was necessary to determine values of the variables between the tabular points. These intermediate values were obtained by Sterling interpolation [11] truncated with second differences. The fifty equations for $\mathbf{P}(t)$ and the $\mathbf{H}_j(t)$ were integrated simultaneously, requiring only one evaluation of $g_i(t)$ and the Jacobian matrix at each tabular value of time. After the integration was terminated at the final time T, the combination coefficients C_j were evaluated from Eq. (8.40), and the stored values substituted into Eq. (8.41) to obtain the new approximation: $\mathbf{X}^{(n+1)}(t)$.

Computation was saved by relaxing truncation error requirements where possible. At the beginning of each iteration the best available initial conditions were used for $\mathbf{P}(t)$. The contributions of the $\mathbf{H}_j(t)$ to $\mathbf{X}^{(n+1)}(t)$ therefore diminished as the process converged. Thus, the accuracies of the $\mathbf{H}_j(t)$ were not as important as that of $\mathbf{P}(t)$.

Convergence of the quasilinearization process was tested by comparing the values of $\mathbf{X}^{(n)}(t)$ and $\mathbf{X}^{(n+1)}(t)$ at the storage interval according to Eq. (8.42). The entire process described above was performed by a general purpose quasilinearization subroutine that utilized double-precision FORTRAN IV (see Chapter 9).

8.4.2 Initial Conditions

The extreme sensitivity of the orbital transfer maneuvers considered here was promptly discovered. For instance, small changes in the durations of the thrust periods were found to produce large variations in the trajectory. Such small variations would often cause the process to diverge. It was therefore necessary to obtain a realistic approximation to the optimal trajectory prior to the initiation of the quasilinearization computations.

It was found that the optimum two-impulse orbital transfer yielded an excellent initial approximation to its finite-thrust counterpart. Accordingly, an initial conditions subroutine, which utilized a steep descent numerical optimization program ([8] and [12]), was employed to generate an initial approximation to the trajectory $\mathbf{X}^{(0)}(t)$. This procedure yielded excellent time histories for the state variables r, ϕ, ρ, y, and m. Utilizing Eq. (8.1), and noting the impulses (velocity change) given by the two-impulse program, it was possible to predict the duration of each "burn period" with accuracy. (Note that one first obtains the change in mass associated with the burn period, and that one must specify a set of rocket parameters, e.g., specific impulse, exhaust velocity, mass flow rate, etc.) In order to generate the initial conditions, it was assumed that the impulsive take-off and arrival points occurred at the center of each burning period.

To avoid having a thrust initiation or termination point occur at $t = 0$ or $t = T$, a coasting arc of several hundred seconds duration was assumed to occur at the beginning and end of the final optimal trajectory. Under this assumption it was a straightforward matter to compute exact initial and final values of the first four state variables. An initial approximation to the total time T required for the maneuver was obtained by summing the transfer times corresponding to the three impulsive coasting arcs.

Although the initial shape of the switching function k was unknown, the impulsive solution gave an excellent approximation to the "switch points" (i.e., $k = 0$). In prior arguments it was established that only these critical values of the switching function are required for the generation of $\mathbf{X}^{(n+1)}(t)$. One may verify this by referring to Eqs. (8.53) through (8.58) and by noting that the Jacobian matrix contains no partial derivatives with respect to k except at a switching point. Note, however, that Eq. (8.53) also requires an initial guess as to the value of k' at the switching points.

Having determined the time histories of the state variables and the switching times, it is also necessary to supply initial approximations to the time histories of the Lagrange multipliers. As previously noted, one of the Lagrange multipliers can be employed as a scale factor. Therefore, λ_1 was assigned an arbitrary initial value. The following relationship may then be constructed from Eq. (8.31) by assuming that $t = 0$ at the impulsive switch point (see McCue and Bender [7]):

$$\lambda_6 = \frac{-\rho\lambda_1 - y\lambda_3}{\rho'/(r \tan v) + y'} \tag{8.59}$$

The impulsive solution may be employed to determine approximate values of r, ρ, y, ρ', y' and v at the center of the initial burning period. By presuming an initial value for λ_3, one may then extract the resulting initial value for λ_6. An initial value for λ_4 then may be computed from the following expression, which is a consequence of Eqs. (8.24) and (8.25):

$$\lambda_4 = \lambda_6/r \tan v \tag{8.60}$$

Thus, the initial values of the Lagrange multipliers can be established by utilizing the impulsive solution and guessing the ratio: λ_1/λ_3. Time histories of the Lagrange multipliers were then produced by an integration procedure that employed Eqs. (8.15) through (8.18) and utilized the previously stored impulsive time histories of the state variables. Figure 8.2 illustrates the validity of this procedure for a typical computer run by comparing the initial and converged time histories of several Lagrange multipliers. The initial approximations to the state variables were considerably better and

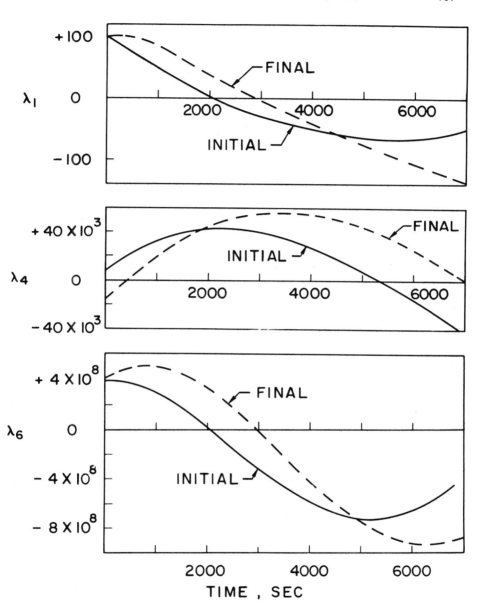

Figure 8.2.   Comparison of Initial and Converged Time Histories of
Several Lagrange Multipliers.

in most cases differences between initial approximations and final converged values could not be detected when plotted to the scale of Fig. 8.2. The fact that the state variables were initially well defined was important to achieving convergence by the quasilinearization technique.

8.4.3 Variable Length Storage Table

Although straightforward application of quasilinearization will result in a solution of the problems considered here, it was found necessary to employ a number of refinements to assure accuracy and proper convergence. For instance, when storing the tabular values of the solution $X^{(n)}(t)$ for use in producing $X^{(n+1)}(t)$, it was necessary to provide a method of maintaining integration accuracy over all portions of the trajectory. That is, it was found necessary to increase the data point storage density during the burning periods. For this reason, the quasilinearization process was constructed about a storage table having variable storage intervals. Figure 8.3a illustrates this concept. The basic quasilinearization subroutine described in Chapter 9 was programmed to integrate over each segment of

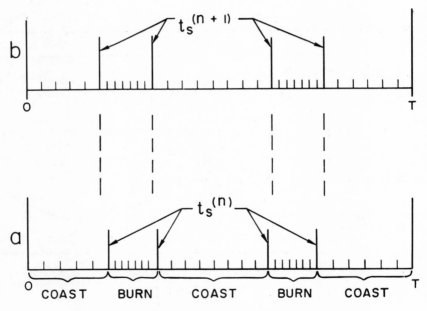

Figure 8.3.   Variable Length Storage Table (a)
and Table with Switching Points t_s Shifted for Next Iteration (b).

the table and to stop at the boundaries. At this point a new storage interval would be introduced and the process continued.

8.4.5 Shifting Storage Table

Because the "bang-bang" control problem is inherently discontinuous at the corner points, significant numerical problems are encountered. At such points one enters a new flight regime wherein $X^{(n)}(t)$ has entirely different properties. For this reason, the performance of valid numerical interpolation across a switching point is not normally possible.

As the quasilinearization process converges, the switching times indicated by the n'th iteration will not coincide with those given by the $n+1$'st iteration. If the switching point occurs between two table entries it is necessary to perform forward interpolation to arrive at appropriate stopping conditions and backward interpolation to arrive at appropriate numerical values to restart the integration. Since two different flight regimes are involved, one finds that the interpolated values at the switching point, in general, do not agree. In many instances the above problem results in serious numerical errors or convergence failure.

One may employ another "trick" to insure numerical integration accuracy at the corner points. After each iteration of quasilinearization new switching times are determined that, in general, will not occur at the previous tabular values. A new table may then be constructed wherein the new switching times are used to define the boundaries of each variable length storage array. This makes it possible to perform a new iteration of quasilinearization with the assurance that each stopping point coincides with a table entry. This new table and its relationship with that used in the prior iteration is illustrated in Fig. 8.3b. Note that this method eliminates the necessity of interpolating to obtain the values at the stopping points. This numerical continuity across the corner points was found to be essential for the accurate convergence of the orbital transfers considered here.

8.4.6 Switch Point Analysis

As was pointed out earlier, the "bang-bang" control process produces trajectories that are very sensitive to the initiation, termination, and duration of thrusting periods. It was, therefore, necessary to employ a rather sophisticated process for determining and controlling the switching times to be utilized in the determination of $X^{(n+1)}(t)$.

A numerical procedure for determining the zeroes of the switching function was employed at the end of each quasilinearization iteration. If the new switching times showed large deviations from the previously used values, the new times *would not* be adopted. Instead, the program would shift the thrust initiation and termination times by a small portion of the indicated change.

Newly determined switching times would be fully utilized only when close agreement with the previous iteration had been achieved. In general such close agreement could only be expected to occur after the quasi-linearization process had proceeded through several iterations. However, once this requirement was met, the program was completely free to use the switching times indicated by the n'th iteration during the computation of the $n+1$'st. The above programmed constraints forced the solution to conform to impulsive initial conditions until the process had achieved sufficient convergence to adequately control itself. Without this constraint and without the judicious use of the impulsive initial conditions, it was usually impossible to obtain convergence.

8.5 NUMERICAL RESULTS

The forementioned IBM 7094 double precision program was utilized to generate transfers between arbitrary coplanar noncoapsidal orbits. Numerical results are best described by comparing the optimal finite-thrust solutions with corresponding optimal impulsive transfers.

8.5.1 Control Variables

As was previously noted, the state variable time histories produced from the impulsive solution showed excellent agreement with the corresponding values for the finite-thrust maneuver. Similar agreement was found for the control variable. This is best illustrated by an example. Initial, final, and transfer orbits corresponding to an optimal finite-thrust transfer are depicted in Fig. 8.4.† The orbit and vehicle parameters are as follows: $p_1 = 5000$ mi., $p_2 = 6000$ mi., $e_1 = e_2 = 0.2$, $\omega_1 = -90°$, $\omega_2 = +30°$, $\beta = 0.0001\ m_0/\text{sec}$ and initial $F/W = 0.4$. Figure 8.4 also indicates the directions and relative magnitudes of the two "impulses" (F_1 and F_2). The transfer orbit of the optimal finite-thrust transfer coincides with its impulsive counterpart when plotted to the scale of Fig. 8.4. The small arcs over which the engine is burning are also noted.

---

† Figure 8.4 is used for illustration purposes only. Actually, it depicts a slightly different optimum transfer between two orbits that are inclined 5°.

Figure 8.5 presents a time history of steering angle v for the orbit transfer maneuver depicted in Fig. 8.4. Note that only a small portion of the steering angle curve has physical significance. These two portions of the curve are expanded in the inset diagrams in Figure 8.5. The inset diagrams also give the impulsive steering angle, for comparison with the finite-thrust solution. For this intermediate thrust case, the thrust initiation and termination times derived from the impulsive solution differed by only a few seconds from those indicated by the quasilinearization solution.

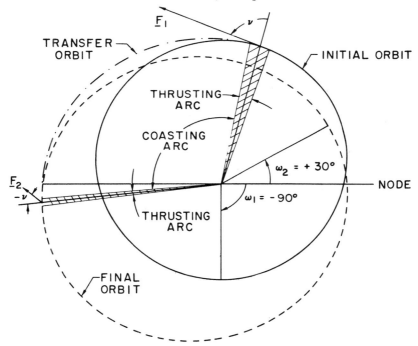

Figure 8.4. Geometry of an Optimal Finite-Thrust Transfer Maneuver (Thrust Vector and Steering Angles for an Impulsive Transfer).

The switching function for this maneuver appears as Fig. 8.6. It is clear that the engine is burning for a small portion of the total time required for the orbital transfer. In order to achieve adequate convergence of the quasilinearization process, the switching times must be determined to approximately 0.001 sec. Referring to Fig. 8.7 and noting that the maneuver may involve several thousand seconds, one obtains an appreciation for the accuracy that must be maintained. Furthermore, when one considers that this accuracy must be maintained during the computations inherent in Eq. (8.21), it is clear that double precision arithmetic is a necessity.

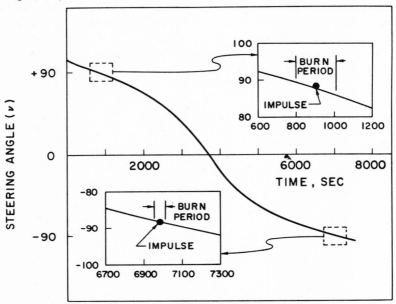

Figure 8.5.   Steering Angle History Compared to Corresponding Impulsive Thrust Angles.

8.5.2   Convergence and Validity Tests

Several tests of a solution's convergence and validity are available. First, because the best approximation to $\mathbf{P}(0)$ is utilized for each iteration, the C_j appearing in Eq. (8.40) should approach zero as the process converges. This provides the first test of convergence. Another test of convergence may be performed by noting successive values of the ξ_i given by Eq. (8.42). During successive iterations, the ξ_i should also approach zero. One should also observe the switching times approaching appropriate constant values during successive iterations. Also, since k must be zero at each switch point one should observe appropriate tabular values of k becoming successively smaller with each iteration. All of the above convergence criteria were achieved during the computation of the solutions presented here.

One may compute certain constants of motion as a test of the solution's validity, accuracy, etc. For instance, the Hamiltonian, which is given by Eq. (8.27), must remain constant. Of course, it is also possible to compute energy and angular momentum along the coasting arcs and to note if these constants are really constant. For all of the results presented here, solutions

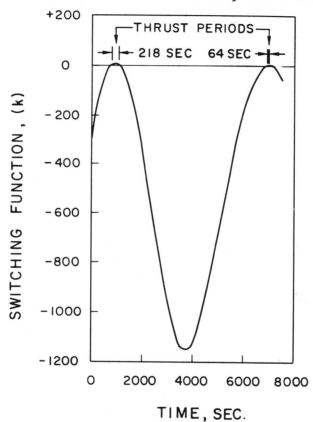

Figure 8.6.   Switching Function Time History.

wherein energy and angular momentum remained constant to more than eight significant figures were achieved. Similar accuracy was achieved for the Hamiltonian. Although this extreme accuracy would not ordinarily be necessary for engineering studies, it was required for the accurate comparison of finite-thrust maneuvers and their impulsive counterparts.

As a final test, the converged solution $\mathbf{X}^{(\infty)}(0)$ was utilized as initial conditions for a straightforward integration of the Euler-Lagrange differential equations. The solution produced by this method was then compared with that generated by quasilinearization. In all cases this test confirmed the validity of the quasilinearization solution.

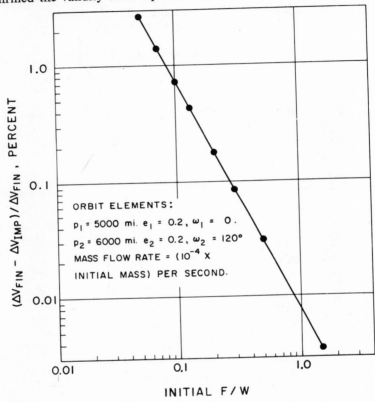

ORBIT ELEMENTS:

$p_1 = 5000$ mi. $e_1 = 0.2$, $\omega_1 = 0$.

$p_2 = 6000$ mi. $e_2 = 0.2$, $\omega_2 = 120°$

MASS FLOW RATE $= (10^{-4} \times$

INITIAL MASS) PER SECOND.

Figure 8.7. Difference in ΔV Required for Finite-Thrust and Impulsive Transfers Versus Initial Thrust-to-Weight Ratio; Orbit Elements: $p_1 = 5000$ mi., $e_1 = 0.2$, $\omega_1 = 0$; $P_2 = 6000$ mi, $e_2 = 0.2$, $\omega_2 = 120°$; Mass Flow Rate $= 10^{-4} \times$ Initial Mass per Second.

8.5.3 Minimum Fuel With Final Time Open

Because the transfer time T derived from the impulsive solution is slightly nonoptimal, additional computations must be performed to determine that trajectory which is time-optimal as well as fuel-optimal. Since the Hamiltonian [Eq. (8.27)] may be thought of as the partial derivative of final mass with respect to T it is necessary to adjust T until the Hamiltonian approaches zero. This was accomplished by perturbing T and computing an additional fuel-optimal trajectory. This required several additional quasilinearization iterations. The Hamiltonian corresponding to each of the fuel-optimal trajectories was examined and simple linear extrapolation was used to predict a new value of T corresponding to $A = 0$. Thus, it was possible to compute a trajectory for which T was "locally" optimal.

8.5.4   V Requirements (Impulsive Thrust Versus Finite-Thrust)

Numerous optimal trajectories were computed in order to produce a comparison of optimal impulsive transfers and corresponding finite-thrust maneuvers. Figure 8.7 compares the velocity change ΔV required for finite-thrust and two-impulse maneuvers over a wide range of initial thrust-to-weight ratios F/W. It was produced by beginning with the transfer maneuver depicted in Fig. 8.5 and parametrically varying the specific impulse (note that β was held constant). The original log-log plot of this data showed no deviation from a straight line (parabola) over the range shown. The fact that the impulsive orbital transfer is a very close approximation to the finite-thrust maneuver is verified by the small percentage differences in Fig. 8.7.

Another interesting comparison was produced by varying the relative perigee angle $\Delta\omega$ of the two coplanar elliptical orbits of Fig. 8.5. Figure 8.8 demonstrates that the difference in velocity change required for the impulsive and the finite-thrust maneuvers exhibits a strong dependence upon $\Delta\omega$. The curve is divided into two regimes corresponding to intersecting and nonintersecting orbits. Near the value of $\Delta\omega$, which corresponds to tangency ($\Delta\omega = 53 \cdot 1301°$), the curve abruptly but continuously changes.

This particular phenomenon is best explained by reference to Fig. 8.9, which contains curves for the two-impulse and finite-thrust maneuvers as separate plots. Figure 8.9 concerns a small range of $\Delta\omega$ over which the orbits are "almost tangent". Since it was known that the class of "almost tangent" orbits produced a number of interesting results ([6] and [7]), considerable effort was devoted to accuracy in examining these critical orientations. Note that the two curves are nearly coincident as long as the orbits do not intersect and that the curves become separate as intersection deepens. This sudden diverging of the two curves in Fig. 8.9 explains the abrupt change noted in Fig. 8.8. The magnitude of this difference depends

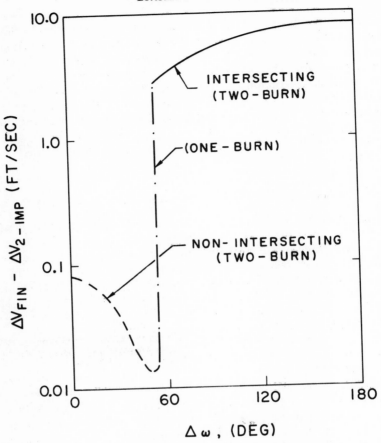

Figure 8.8. ΔV **Difference as a Function of the Relative Perigee Orientation of Two Elliptical Orbits.**

upon the particular rocket parameters employed (e.g., specific impulses, mass flow rate, etc.).

Contensou [13], Lawden [14], and Breakwell [15] discuss the existence of an optimal one-impulse maneuver. The two-impulse curve shown in Fig. 8.9 contains a small region over which a one-impulse transfer between the two orbits is optimal. The finite-thrust curve is composed of a number of points that are designated one-burn maneuvers or two-burn maneuvers. Note that an optimal one-burn maneuver also exists over a small range

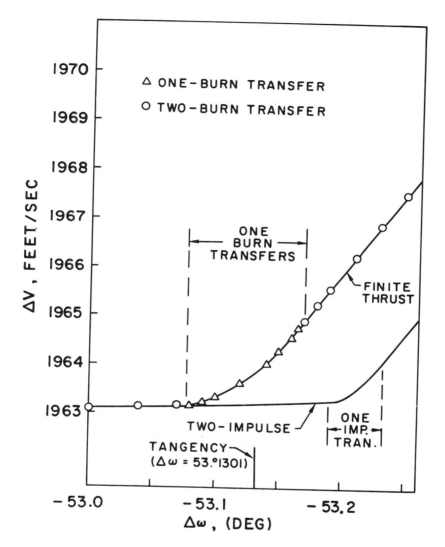

Figure 8.9.   Velocity Change Required for Finite-Thrust and
Impulsive Transfers Between "Almost Tangent" Orbits.
△ One-burn Transfer and ° Two-burn Transfer.

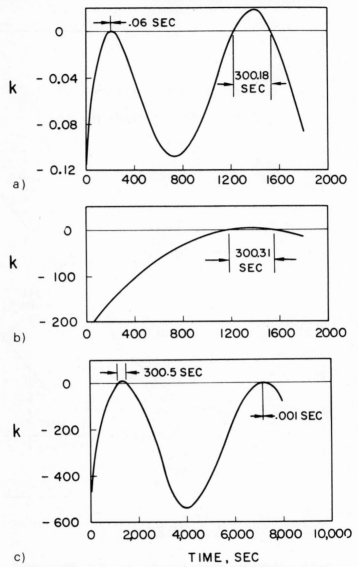

Figure 8.10.   Switching Function Time Histories for Several
Transfers Between "Almost Tangent" Orbits;
a, Nonintersecting, Two-burn Transfers ($\Delta\omega = -53.07°$),
b, Nonintersecting, One-burn Transfers ($\Delta\omega = -53.12°$), and c, Shallowly
Intersecting, Two-burn Transfers ($\Delta\omega = -53.17°$).

of $\Delta\omega$. Thus, one obtains conditions that are again analogous to those found for impulsive transfer.

As in the impulsive case, the maneuvers represented by points directly on either side of the one-burn region are characterized by entirely different steering angle and thrust time histories. To the left of the one-burn region, the first burn period is rather small and the thrust for both burns is in the forward direction. To the right of the one-burn region, the second burn period is very small and its thrust direction opposes the vehicle's velocity vector. Figure 8.10 presents the switching function time histories associated with each of three kinds of finite-thrust maneuvers considered in Fig. 8.9. For the two-burn transfers, the duration of the smaller burn periods was a fraction of a second as compared to about 300 sec for the larger burn period. Some solutions that were near the boundary of the one-burn region exhibited a second burn period of 0.002 sec duration and smaller. The extreme numerical accuracy required to produce the results of Figs. 8.8 and 8.9 is evident.

8.6 CONCLUSIONS

The numerical results demonstrate that the quasilinearization technique can be a powerful tool for the optimization of "bang-bang" control problems. However, it appears that a good first approximation to the solution is required to insure convergence. In general it appears that impulsive orbital transfer maneuvers provide an excellent approximation to their finite-thrust counterparts.

NOMENCLATURE

A first integral
B boundary condition vector
C_j combination coefficients required for quasilinearization process
c effective exhaust velocity
D auxiliary variable defined by Eq. (8.26)
e eccentricity
F thrust

G quantity to be minimized (maximized)
g time derivative vector
H_j homogeneous solution vectors required for quasilinearization process
k switching function
m mass
N number of differential equations

n	quasilinearization iteration number	$\mathbf{X}$	dependent variable vector
$\mathbf{P}$	particular solution vector required for quasilinearization process	X_i	components of vector $\mathbf{X}$
		y	angular velocity $\cdot$
		β	mass flow rate
p	semi-latus rectum	λ_i	Lagrange multiplier
r	radius	μ	gravitational constant
T	total time required for transfer maneuver	ν	steering angle
		ρ	radial velocity
t	time	ϕ	central angle
		ω	argument of perigee

REFERENCES

1. S. A. Jurovics, "Orbital Transfer by Optimum Thrust Direction and Duration," North Am. Aviation, Inc. SID 64-29 (February, 1964).
2. R. Kalaba, "Some Aspects of Quasilinearization (Nonlinear Differential Equations and Nonlinear Mechanics)", pp. 135–146. Academic Press, New York, 1963.
3. R. Bellman, H. Kagiwada, and R. Kalaba, "Quasilinearization, System Identification and Prediction." RAND Corp., RM-3812 PR (August, 1963).
4. R. McGill and P. Kenneth, Solution of variational problems by means of a generalized Newton–Raphson operator. *AIAA J.* **2**, 1761-1766 (October, 1964).
5. G. A. McCue, Optimum two-impulse orbital transfer and rendezvous between inclined elliptical orbits. *AIAA J.* **1**, 1865–1872 (1963).
6. G. A. McCue, Optimization and visualization of functions. *AIAA J.* **2**, 99–100 (January, 1964).
7. G. A. McCue and D. F. Bender, "Optimum Transfer between Nearly Tangent Orbits," North Am. Aviation, Inc., SID 64-1097 (May, 1965).
8. G. A. McCue and D. F. Bender, Numerical investigation of minimum impulse orbital transfer. *AIAA J.* **3**, 2328–2334 (December, 1965).
9. G. Leitmann, Variational problems with bounded control variables *in* "Optimization Techniques," (G. Leitmann, ed.), pp. 171–204. Academic Press, New York, 1962.
10. B. Friedman, "Principles and Techniques of Applied Mathematics," Wiley, New York, 1962.
11. F. B. Hildebrand, "Methods of Applied Mathematics," pp. 34–35. Prentice-Hall, Englewood Cliffs, New Jersey, 1952.
12. G. A. McCue and R. C. Hoy, "Optimum Two-Impulse Orbital Transfer Program", North Am. Aviation, Inc., SID 65-1119 (August, 1965).
13. P. Contensou, Étude théorique des trajectoires optimales dans un champ de gravitation. Application au cas d'un centre d'attraction unique. *Astronaut, Acta* **8**, 134–150 (1963); also Gruman Res. Dept. Transl. TR-22 (P. Kenneth, transl.) (August, 1962).
14. D. F. Lawden, "Optimal Trajectories for Space Navigation," Butterworth, London and Washington, D.C., 1963.
15. J. V. Breakwell, "Minimum Impulse Transfer", AIAA Preprint 63-416 (1963).

Chapter 9

A GENERALIZED SUBROUTINE FOR SOLVING QUASILINEARIZATION PROBLEMS

9.1 INTRODUCTION

This chapter describes the authors' basic set of quasilinearization programs and their use in solving the simple problem in Chapter 3. Most of the computational details of quasilinearization that are presented here and in Appendix 1 are common to large classes of problems and need be programmed only once. As a result, readers who wish to adapt the authors' routines to solve their particular problem will find their task simplified.

9.2 PROGRAMMING PHILOSOPHY

In the process of obtaining solutions to the problems described in Chapters 3 through 8, the authors developed a generalized quasilinearization FORTRAN computer program in the form of a major subroutine and associated lower level subroutines. This set of subroutines contains the facilities to handle the computations for all of the classes of problems in the foregoing chapters[†] when the information that is peculiar to the problem is furnished in the specified manner. In order to meet the diverse requirements of these problems, the program contains a number of sections that are used only as required by the special features of the problems. For example, the orthogonalization of homogeneous solutions is used in most of the hydrodynamic problems but not in the astrodynamic problem, whereas only the astrodynamic problem requires the special stopping feature in the integration.

This set of programs is not claimed to be optimal in any sense but is presented merely as a working example, which may aid the reader in understanding and implementing the ideas and methods presented in the first eight chapters. For example, integration and interpolation routines can be upgraded to incorporate the latest advances in many ways. Compromises

---

[†] The problem in Section 5.11 requires additional coding.

between efficiency and stability should be made for each separate class of problems if a large number of solutions are to be computed.

The line of descent of the present program has been from FORTRAN II to FORTRAN IV for the IBM 7090–7094 computers and more recently to G and H level FORTRAN for the IBM 360 computer.[†] Due to this derivation the program uses very few of the special features of the third-generation software, although these features have not been avoided entirely in the process of conversion. Some improvement in the conciseness and efficiency of the code should be obtained with full use of the new software features.

Assembly language coding has been scrupulously avoided in the development of these programs except where such coding was found in library programs, such as the simultaneous equation subroutine. The expected advent of third-generation hardware was, to a large degree, responsible for this decision to avoid machine-dependent coding. Regardless of what the projected date of the $(n+1)$st generation may be, the rate of obsolescence of certain numerical procedures is sufficiently high, so that assembly language coding still does not appear advantageous.

In this chapter our plan of attack is to describe the general-purpose quasilinearization subroutine in some detail and then mention briefly the lower level subroutines. We shall conclude with a discussion of the main program and associated specific subroutines for a simple example—the boundary-layer problem in Chapter 3.

9.3 QASLN, A GENERAL PURPOSE QUASILINEARIZATION SUBROUTINE

Most of the logic and the computations associated with the application of the quasilinearization algorithm have been placed in a subroutine called QASLN. There are several reasons for setting up this program as a subroutine rather than a main program, although it does contain the principal part of the code. First, the inputs that specify the problem may be either read in or generated by lengthy separate computations as is the case with the problem in Chapter 8. Second, the variable dimension feature of FORTRAN IV and FORTRAN G and H is utilized by specifying fixed dimensions of large arrays only in the main program or a higher level subroutine that is short and can be economically recompiled. This permits the available storage to be efficiently allocated and/or changed to suit a particular problem. In some

---

[†] The standard configuration for the North American Rockwell Corporation's Aerospace and Systems Group is an IBM 360 Model 165.

other high-level language such as PL/1, the storage can be allocated dynamically, relieving the necessity for this type of variable dimensioning.

A concept of the relation of QASLN to the other parts of the quasi-linearization program package can be gained by reference to the modular tree structure in Fig. 9.1. Although it is not the main program, QASLN is the principal program in that it contains all of the logic and computations that are peculiar to quasilinearization. The other modular parts of the tree are of four types: (1) the main program and certain subroutines that

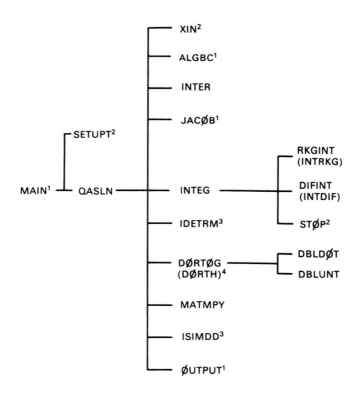

Figure 9.1. Modular Tree Structure of Quasilinearization Program Package.

always contain information peculiar to the particular problem, such as the Jacobian matrix and the output format; (2) those subroutines that are useful in a large class of problems, but may require replacement for certain complex problems such as that of Chapter 8; (3) the library subroutines, such as the simultaneous equation solver ISIMDD, which should be available at any facility; and (4) those subprograms, such as the integration routines that perform a function not peculiar to quasilinearization but that are specially adapted to couple with the structure of QASLN.

The modular tree structure chart in Fig. 9.1 is supplemented by Table I, which lists the modules in the tree with short descriptions of their functions in the program. More detailed information on these modules is provided

Table I.   Functions of Program Modules

Module Name	Module Function
MAIN	Main program: allocates storage, reads control data (e.g., boundary conditions, convergence criteria) for quasilinearization; reads data for specific problem (e.g., wedge angle β)
SETUPT	Constructs independent variable table
QASLN	Quasilinearization: performs logical operations and computations to implement the quasilinearization algorithm
XIN	Reads data and constructs polygonal initial approximation to solution
ALGBC	Computes initial conditions on variables that are dependent on other variables or parameters and Jacobian matrix of these conditions. (This is a dummy subroutine in this example.)
INTER	Interpolates for state vector in table of last solution
JACØB	Computes derivative of state vector and Jacobian matrix based on last approximation to solution
INTEG	Controls integration: starting the difference integration with Runge-Kutta integration, stopping at tabular points and special stopping procedure at prescribed points
INTRKG	Runge-Kutta-Gill integration: starts integration for INTDIF
INTDIF	Difference integration: builds table integrates with Adams-Moulton formula
STØP	Carries out special procedure at stopping points (e.g., makes step changes in variables, as in Chapter 8). (Dummy subroutine in this example.)
IDETRM	Evaluates determinant (used in calculating "condition" of homogeneous solution matrix.)
DØRTH	Orthogonalizes solution vectors in homogeneous solution matrix
DBLDØT	Forms dot product of two vectors
DBLUNT	Calculates length and associated unit vector for a given vector
MATMPY	Performs matrix multiplication
ISIMDD	Solves set of linear equations
ØUTPUT	Converts converged solution to final form and prints solution in convenient format with identifying information

by parameter tables, flow charts, discussed below, and the listings of the programs that are found in the appendix.

Subroutine parameter tables, Tables III through XIII, are presented for most of the nonlibrary subroutines to show how information is communicated between the program modules. Only variables that pass in and/or out of the subroutine are shown in these tables. However, this restriction eliminates few variables of any importance. The first column of these tables gives the FORTRAN name of each variable. Dimensions of the variables, which may be fixed or computed at each entry are given in the second column. The third column indicates whether the variable is communicated by call list or common block, although in a few cases the subroutines read from or write to external data sets. Whether the variable is input I, output Ø, or both input and output is indicated in the fourth column. Finally, the fifth column gives a short description of the variable.

Table II. QASLN Symbol Table

FORTRAN Symbol	Engr. Symbol	Type and Length	Dimension Common	Use	Description
ACUM		Real*8	(NU, NS)	C	Temporary storage
B	**B**	Real*8	(10) /ZØØ/	I	Boundary condition vector: $B(I)$ is boundary condition on $X(M(I))$ at VARIND $(L(I))$
C		Real*8	(5)	C	Combination coefficient vector used in inverse transformations
CS		Real*8		C	Reduction in length to make $P2$ a unit vector
CTP	^kC_T	Real*8	(N2, NTM)	C	Coefficient vector for scaling P by subtracting $H \cdot c$
D	$\delta X^{(k)}$	Real*8	(20)	S	First central difference of state vector: used only by INTER
DTPRNT	t	Real*8	/ZI/	C, Ø	Spacing of tabular points in a segment of the solution
DXKP1	$dX^{(k+1)}/dt$	Real*8	(NU, NS)	C, Ø	Current values of derivative of state vector for all solutions (P and H)
E	e	Real*8	(6, 6) /ZT/	I	Matrix of unit vectors of ortho-normalized subspace of homogeneous solutions
EPSP	ε	Real*4	/ZX/	I	Minimum allowable condition number for homogeneous solution subspace matrix $H2$

Table II.   QASLN Symbol Table (continued)

FORTRAN Symbol	Engr. Symbol	Type and Length	Dimension Common	Use	Description
F		Real*8		C, Ø	Scale factor for determinant in solution of linear equations
G	g, F	Real*8	(20)	I	Derivative of state vector or computed initial condition vector, respectively
H1	H	Real*4	(N1, N2, IMAX)	C	Homogeneous solutions matrix (stored as a function of independent variable)
H2	$\hat{H}$	Real*8	(6, 6)	C	Matrix of homogeneous solution components corresponding to noninitial boundary conditions
H3		Real*8	(6, 8) /ZT/	C	Matrix of values of particular solution components at the locations where they must satisfy boundary conditions
IA		Real*8	(20) /ZX/	C	Combination coefficient vector for homogeneous solutions. Also temporary storage
IIT		Integer*4		C	Control for subroutine ØUTPUT, which indicates where in QASLN it was called
IMAX		Integer*4		I	Number of storage points for solution (as function of independent variable); also location in table where final value is stored
I1		Integer*2		C	Index of tabular point at beginning of segment of solution
I2		Integer*2		C	Index of tabular point at end of segment of solution
KP		Integer*4		C	Index number of segment of solution (tabular interval is constant within a segment)
KSUB		Integer*4	(10) /ZSUB/	I	Index numbers of tabular points of solutions at boundaries of segments (tabular interval is constant within a segment)
KTRANS		Integer*4	(50) /ZT/	I	Sequence numbers of solution tabular points at which orthonormalizing transformations were made

Table II.   QASLN Symbol Table (continued)

FORTRAN Symbol	Engr. Symbol	Type and Length	Dimension Common	Use	Description
L		Integer*2		C	Index counted from end of interval used in summing particular and homogeneous solution to obtain $(k+1)$st approximation
MEMBER	M	Integer*4	(10) /zøø/	I	Vector whose nth integer component specifies to which component of the state vector the nth boundary condition applies
MI		Integer*4	(10) /zøø/	I	Sequence numbers of elements of state vector that are varied at initial point
MT		Integer*4		C	Number of orthonormalizing transformations made up to current value of independent variable
MU	$\mu X^{(k)}$	Real*8	(20)	S	Mean value of state vector between two tabular points $\mu X_n = \frac{1}{2}(X_{n-\frac{1}{2}} + X_{n+\frac{1}{2}})$
MUD2	$\mu \delta^2 X^{(k)}$	Real*8	(20)	S	Mean second difference of state vector
MX		Integer*4		I	Error indicator returned by determinant evaluation or simultaneous equation-solution function subroutines
NI		Integer*4	/zi/	C, I, Ø	Index of integration substeps (Runge-Kutta fourth-order integration has four substeps; difference integration has two substeps)
NITER		Integer*4		C, Ø	Current iteration number in quasilinearization process
NMAX		Integer*4	/zi/	C, Ø	Index of tabular point at upper boundary of a segment of the solution
NMIN		Integer*4	/zi/	C, Ø	Index of tabular point at lower boundary of a segment of the solution
NRST		Integer*4	/zi/	C, I, Ø	"Restart" index for integration —the number of integration steps since integration was last started (NRST = -1 when restarted)

Table II. QASLN Symbol Table (continued)

FORTRAN Symbol	Engr. Symbol	Type and Length	Dimension Common	Use	Description
NC		Integer*4		C	Number of parameters (constants) in state vector
NS		Integer*4		I	Total number of solutions (particular plus homogeneous)
NTM		Integer*4		I	Maximum number of orthogonalizations allowed (dimension of storage)
NU		Integer*4		I	Total dimension of state vector (variable plus parameters)
NU1		Integer*4		C	Number of components of the state vector that appear in the equations (NBP components appear only in the boundary conditions)
N1	n	Integer*4		I, Ø	Number of *variables* in state vector [also $N(1)$]
N2	M	Integer*4		I, Ø	Number of boundary conditions that are not initial conditions [also $N(2)$]
N(4)		Integer*4	/ZØØ/	I	Print control for use in checking new programs
N3		Integer*4		C	Maximum number of iterations of quasilinearization allowed before process is terminated [also $N(3)$]
N5	n_5	Integer*4		C	Number of noninitial boundary conditions. If $N5 > N2$, the boundary conditions are satisfied in a least-square sense
P	P	Real*8		I, C, Ø	Particular solution vector as a function of independent variable (N1, IMAX)
P2		Real*8	(6) /ZT/	C	Vector composed of those components of the particular solution that must satisfy (noninitial) boundary conditions
RHØ		Real*8	(20)	C	Maximum deviations of state vector components over the integration interval between successive approximations $\max_t \left(\|\mathbf{X}^{(k+1)}(t) - \mathbf{X}^{(k)}(t)\|\right)$
RHØS		Real*4	(20) /ZØØ/	I	Convergence criterion for RHØ

Table II. QASLN Symbol Table (continued)

FORTRAN Symbol	Engr. Symbol	Type and Length	Dimension Common	Use	Description
SCALE		Real*4		C	Scale factor for determinant evaluation: inverse of maximum parallelepiped in subspace of homogeneous solutions
STØR		Løgical*1		C	Logical variable, which is true if the current component of the state vector must satisfy a boundary condition at the current point
T	T	Real*8	(N2, N2, NTM)	C	Transformation matrix for orthogonalization
TEMP	$\hat{H}$	Real*4	(12, 12)	C, Ø	Temporary storage matrix for use in computing condition of $\hat{H}$
TNSTØR		Real*8	/ZI/	C, Ø	Value of the independent variable at the next tabular point (storage point)
TX		Real*8	/ZI/	C, I, Ø	Current value of the independent variable
VARIND	t_l	Real*8	(100) /ZSUB/	I	Table of values of independent variable at which solutions are stored
XI		Real*8	(10) /ZX/	I	Nonzero initial values of $H(MI(J),J)$; small variations on the initial conditions
XJAC	$\dfrac{\partial \mathbf{g}}{\partial \mathbf{X}}$ or $\dfrac{\partial \mathbf{F}}{\partial \mathbf{X}}$	Real*8	(N1, NU)	I	Jacobian matrix, for differential equations or initial conditions, respectively
XK	$X^{(k)}$	Real*8	(20) /ZI/	I	State vector interpolated from tabulated previous (kth) approximation at current value of the independent variable
XKP1	$X^{(kH)}$	Real*8	(NU, NS)	I	Current value of the $(k+1)$st approximation to the state vector for homogeneous and particular solutions
XKS	$X^{(k)}$	Real*8	(N1, IMAX)	I, C, Ø	Table of the variable part of the state vector stored as a function of the independent variable

Table III. Parameters for Main Program

Variable Name	Dimension	How Communicated	Input-Output	Description
H1	(3, 1, 101)	Call list	S	H, homogeneous solution as function of n
RHØS	(20)	Read, Common	I, Ø	Solution convergence criterion
XKS	(3, 101)	Call list	S	$X^{(k)}(n)$, previous approximation to solution as function of n
P	(3, 101)	Call list	S	P, particular solution as function of n
JACM	(3, 3)	Call list	S	Jacobian matrix
T	(1, 1, 1)	Call list	S	Orthogonalizing transformations (dummy for this example)
CTP	(1, 1)	Call list	S	Scale factors for P (dummy for this example)
DXKP1	(3, 2)	Call list	S	$d\mathbf{X}^{(k+1)}/dn$ derivatives of P, H
XKP1	(3, 2)	Call list	S	$\mathbf{X}^{(k+1)}$ state vectors (P, H)
ACUM	(3, 2)	Call list	S	Scratch storage
XI	1	Read, Common	I, Ø	Initial value for nonzero element of homogeneous solution ($\Delta f''$)
IA	20	Common	S	Combination coefficient and scratch storage
TSUB	10	Read, Common	I, Ø	Independent variable values at boundaries of segments
B	10	Common	Ø	Vector of noninitial boundary conditions
VARIND	101	Common	S	Independent variable table
BETA	1	Common	I, Ø	Wedge angle (parameter for JACØB)
EPSP	1	Common	S	Orthogonalization criterion (not used in this example)
DATE	1	Call list	I, Ø	Date and execution time (from system subroutines
EXT	1	Written		
NDIF	1	Read, Common	I, Ø	Number of differences used in integration
NBP	1	Common	Ø	Number of parameters in the state vector that appear only in the boundary conditions
N	5	Read, Common	I, Ø	Control constant vector: $N(1) = N1$, number of variables in state vector; $N(2) = N2$, number of homogeneous solutions; $N(3)$, number of iterations allowed for convergence; $N(4)$, print control; $N(5)$, number of noninitial conditions if least-square boundary-condition fitting required [otherwise $N(5) = 0$]

Table III. Parameters for Main Program (continued)

Variable Name	Dimen-sion	How Communicated	Input-Output	Description
NSTP	1	Read, Common	I, Ø	Number of integration steps per storage interval
INTP1	1	Read, Common	I, Ø	Number of segment boundaries in solution (segments plus one)
L	10	Common	Ø	Solution table indices at which boundary conditions are applied
M	10	Common	Ø	Indices of state vector components to which boundary conditions are applied
MI	10	Common	Ø	Indices of homogeneous solution components that have nonzero initial conditions
N1	1	Call list	Ø	See N
N2	1	Call list	Ø	See N
NU	1	Call list	Ø	Total number of components in state vector (variables and parameters)
NS	1	Call list	Ø	Number of solutions to be integrated (N2 homogeneous plus one particular)
NTM	1	Call list	Ø	Maximum number of orthogonalizations
IMAX	1	Call list	Ø	Number of entries in solution table

Table IV. Subroutine Parameters—SETUPT

Variable Name	Dimen-sion	How Communicated	Input-Output	Description
KSUB	10	Common	I	Positions in storage table of boundaries of segments (storage equally spaced within segments)
TSUB	10	Common	I	Values of independent variable at boundaries in storage table
VARIND	100	Common	Ø	Values of independent variable at storage locations
NINTP1	1	Common	I	Number of boundaries of segments (intervals plus one)

Table V. Subroutine Parameters—MATMPY

Variable Name	Dimension	How Communicated	Input-Output	Description
A	$(N1, N2)$ $(N1\ ND1)^{max}$	Call list	I	Left-hand matrix in product
B	$(N2, N3)$ $(N2\ ND2)^{max}$	Call list	I	Right-hand matrix in product
C	$(N1, N3)$ $(N3\ ND3)^{max}$	Call list	Ø	Product matrix
N1	1	Call list	I	Number of rows in A and C
N2	1	Call list	I	Number of columns in A and rows in B
N3	1	Call list	I	Number of columns in B and C
ND1	1	Call list	I	Maximum number of rows in storage for A
ND2	1	Call list	I	Maximum number of rows in storage for B
ND3	1	Call list	I	Maximum number of rows in storage for C

Table VI. Subroutine Parameters—XIN

Variable Name	Dimension	How Communicated	Input-Output	Description
NU	1	Call list	I	Number of variables (rows) in state vector-initial estimate matrix
IMAX	1	Call list	I	Number of storage points (columns) in state vector-initial estimate matrix
Y	$(NU, IMAX)$	Call list	Ø	State vector *variable* initial estimate matrix
N1	1	Read	I	Number of *variables* for which estimate is to be generated
NC	1	Read	I	Number of parameters to be read
M	$N1 \leqslant 15$	Read	I	Numbers of vertices in polygonal approximations for state variables
W	$N1 \leqslant 50$	Read	I	Ordinates of vertices in polygonal approximation for state variables
Z	$N1 \leqslant 50$	Read	I	Abscissas of vertices in polygonal approximation for state variables
VARIND	$IMAX \leqslant 100$	Common	I	Abscissas at which the polygonal approximation is interpolated
XK	$NC \leqslant 20$	Common	Ø	State vector (parameters are output)

Table VII.   Subroutine Parameters—INTER

Variable Name	Dimension	How Communicated	Input-Output	Description
X	(N1, IMD)	Call list	I	Table of vector function f of single independent variable
N1	1	Call list	I	Dimension of vector function
IMD	1	Call list	I	Number of independent variable values in table
MU	N1	Call list	I, Ø	Mean value of X (μX)
D	N1	Call list	I, Ø	First central difference (δX)
MUD2	N1	Call list	I, Ø	Mean second central difference ($\mu\delta^2 X$)
VARIND	IMD	Call list	I	Independent variable table
II	1	Call list	I	Sequence number of largest tabular value of independent variable less than or equal to current value
TX	1	Common	I	Current value of independent variable
IN	1	Common	I	Sequence number of tabular value at beginning of current segment
IM	1	Common	I	Sequence number of tabular value at end of current segment
DT	1	Common	I	Difference of tabulated values of independent variable
XK	N1 ≤ 12	Common	Ø	Interpolated values of components of vector function

Table VIII.   Subroutine Parameters—WEDGJ (JACOB)

Variable Name	Dimension	How Communicated	Input-Output	Description
ETA	1	Call list	I	Independent variable
X	3	Call list	I	State vector (dependent variables)
G	3	Call list	Ø	Vector right-hand side function of differential equations
J	(3, 3)	Call list	Ø	Jacobian matrix
ISW	1	Call list	I	Switch for quasilinearization or final check integration
N(4)	1	Common	I	Extra-print control

Table IX. Subroutine Parameters—OUTPUT

Variable Name	Dimension	How Communicated	Input-Output	Description
X	(N1, IMAX)	Call list	I, Ø	Table: variables in state vector as function of independent variable
N1	1	Call list	I	Number of variable elements in state vector
IMAX	1	Call list	I	Number of independent variable values at which state vector is recorded
IIT	1	Call list	I	Switch for quasilinearization or final check integration
VARIND	101	Common	I, Ø	Values of independent variable
BETA	1	Common	I, Ø	Parameter (wedge angle) in solution

Table X. Subroutine Parameters—INTEG

Variable Name	Dimension	How Communicated	Input-Output	Description
X	(NU, NS)	Call list	I, Ø	Matrix of particular and homogeneous state vectors
DX	(NU, NS)	Call list	I	Matrix of derivatives of particular and homogeneous state vectors
NU	1	Call list	I	Number of variables in state vector
NS	1	Call list	I	Total number of solutions (homogeneous and particular)
ACUM	(NU, NS)	Call list	I, Ø	Storage used by lower level subroutines until TSTØP is reached
TSTØP	1	Call list	I	Value of independent variable at which integration is terminated
T	1	Common	I, Ø	Independent variable
H	1	Common	I, Ø	Integration step size
NI	1	Common	I, Ø	Number of substep
NRST	1	Common	I, Ø	Number of steps since integration was restarted
NDIF	1	Common	I	Number of differences to be used in integration
DT	1	Common	I	Interval on independent variable at which results are stored
XERR	1	Common	I	Allowable error

Table XI.   Subroutine Parameters—INTRKG

Variable Name	Dimension	How Communicated	Input-Output	Description
Y	(NU, NS)	Call list	I, Ø	Integrated variable ($\mathbf{X}$)
K	(NU, NS)	Call list	I	Integrand ($d\mathbf{X}/dt$)
NU	1	Call list	I	Total number of elements in state vector ($\mathbf{X}$)
NS	1	Call list	I	Total number of solutions being integrated (one particular and n homogeneous)
Q	(NU, NS)	Call list	I, Ø	"Memory" of last step
TX1	1	Common	I, Ø	Independent variable
H	1	Common	I	Integration step size
J	1	Common	I, Ø	Integration substep number
N1	1	Common	I	Number of *variables* in state vector $\mathbf{X}$
NRST	1	Common	I	Number of steps since integration was started

Table XII.   Subroutine Parameters—INTDIF

Variable Name	Dimension	How Communicated	Input-Output	Description
Y	(NU, NS)	Call list	I, Ø	Integrated variable ($\mathbf{X}$)
YP	(NU, NS)	Call list	I	Integrand ($d\mathbf{X}/dt$)
NU	1	Call list	I	Total number of elements in state vector ($\mathbf{X}$)
NS	1	Call list	I	Total number of solutions being integrated (one particular and n homogeneous solutions)
TS	(NU, NS)	Call list	I, Ø	Temporary storage for predicted integral
TX	1	Common	I, Ø	Independent variable
H	1	Common	I	Integration step size
NI	1	Common	I, Ø	Integration substep number $NI = 0 \Rightarrow$ predictor, $NI = 1 \Rightarrow$ corrector
NRST	1	Common	I	Number of steps since integration was started
NDIF	1	Common	I	Number of differences retained in integration
DIFTAB	(4, 72)	Common	I, Ø	Table of differences of integrands

Table XIII.   Subroutine Parameters—DORTH

Variable Name	Dimension	How Communicated	Input-Output	Description
A	(NA, N)	Call list	I	Matrix to be orthogonalized
E	(NE, N)	Call list	Ø	Orthonormal matrix derived from A (temporary storage)
L	(N)	Call list	Ø	Lengths of normalized row vectors (temporary storage)
T	(NT, NT)	Call list	Ø	Orthogonalized transformation matrix
BD	(N)	Call list	Ø	Unit vector of orthogonalized matrix (temporary storage only)
NA	1	Call list	I	Maximum row dimension of storage for A
N	1	Call list	I	Dimension of A
NE	1	Call list	I	Maximum row dimension of storage for E
NT	1	Call list	I	Maximum dimension of storage for T

A symbol table (Table II) is given for QASLN that is more elaborate than the parameter tables for the other subroutines. The additional detail is justified by the central importance of QASLN in the program. All important variables used inside QASLN (C in column five) as well as those that are communicated (I, Ø in column five) are given. An S in column five indicates that the variable is merely stored for a lower level subroutine. Where applicable, the engineering symbol for a variable, i.e., the symbol used for it in Chapters 1 through 8, is given in column two. Many of the FORTRAN variables are not simply identified with the symbolic mathematical formulation. The FORTRAN variable type and length (e.g., Real*8) are given in column 3. The double word length for those variables, so indicated, is essential, particularly on a 32-bit word machine. Any integer variable used as a variable dimension in FORTRAN G or H is required to be 4-bytes long, but some of the Integer*4 variables in this program are not so used and could be converted to 2-byte length. Common block name (blank common is not used) is shown below the dimension in the fourth column for those variables transmitted through common.

Flow charts are presented in Figs. 9.2 through 9.10 for the main program and all but the dummy, or the very short subroutines. However, flow charts are not given for library subroutines, which are assumed to be "black

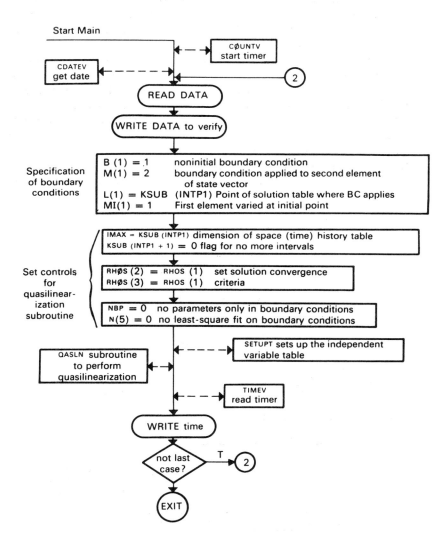

Figure 9.2. Example: Boundary Layer on a Wedge—Main Program.

boxes." The notation used is a compromise between prose description, mathematical formulas, and FORTRAN notation. The authors have felt justified in glossing over some of the detail, particularly in the longer subroutines, since the reader has the option of retrieving the ultimate detail from the listings in the appendix.

In constructing the flow charts, we have used relatively few symbols. Rectangular boxes generally indicate computation and contain description and/or formulas. The start or end of a DØ loop is also indicated by a rectangular box, but with appropriate arrows. Boxes connected by double-headed dashed arrows indicate called subroutines. However, function references or calls to utility subroutines, e.g., MATMPY, may not be shown explicitly. Conditional transfers, either logical or arithmetic IFs, are indicated by diamonds where the type of IF is determined by the number of branches. Remote connections, except at the beginning or end of a page,

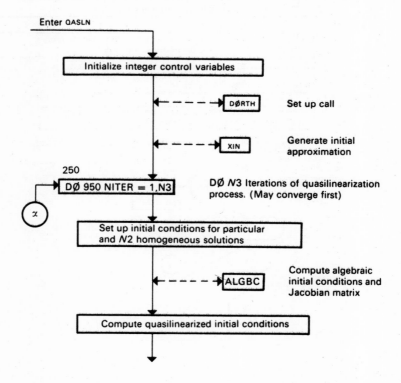

Figure 9.3.　Quasilinearization Subroutine—QASLN.

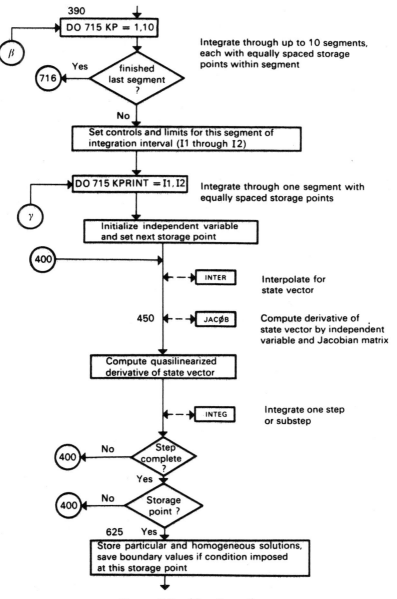

Figure 9.3.   (Continued).

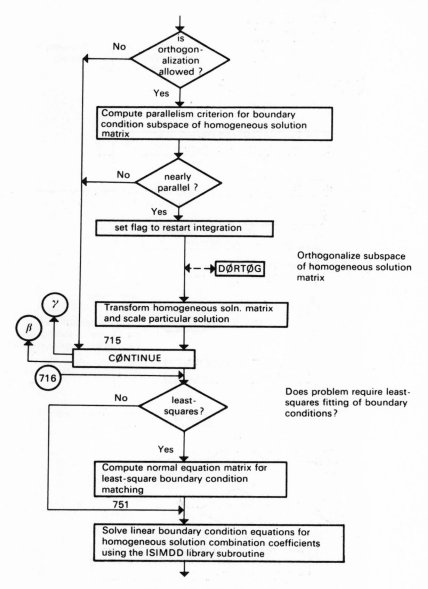

Figure 9.3.   (Continued).

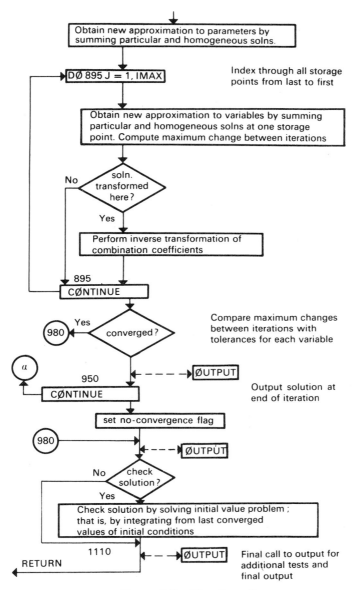

Figure 9.3. (Continued).

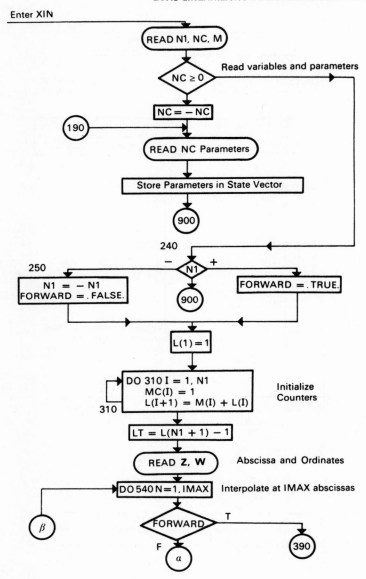

Figure 9.4.   Initial Approximation Subroutine—XIN.

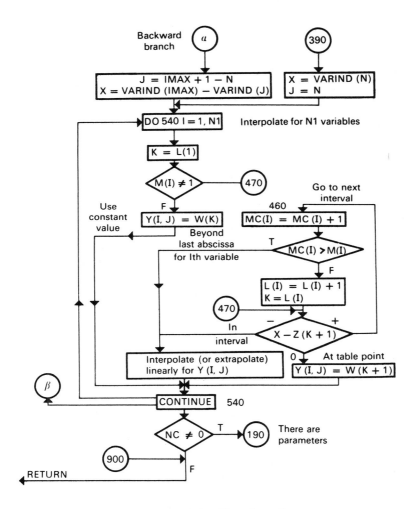

Figure 9.4.   (Continued).

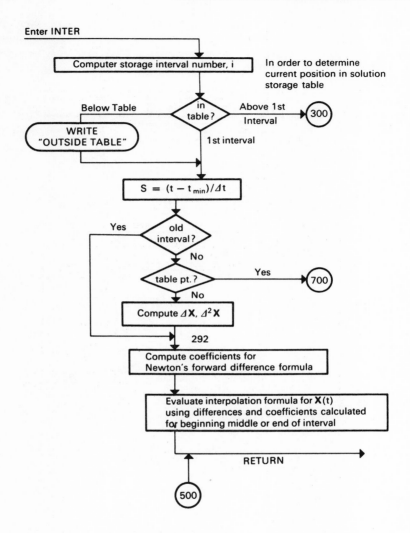

Figure 9.5. Interpolation Subroutine i/m INTER

are denoted by small circles containing a number that refers to a statement number on the corresponding listing, or circles containing Greek letters where the transfer of control is not to a numbered statement. Input-output operations are indicated by boxes with rounded ends and the notation READ or WRITE.

Comments and additional explanation have been placed next to single boxes where additional clarifying information appeared to be valuable or have been added to clarify the function of whole sections of a subroutine.

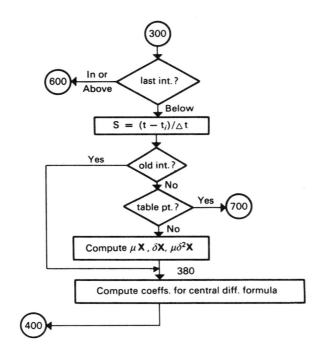

Figure 9.5. (Continued).

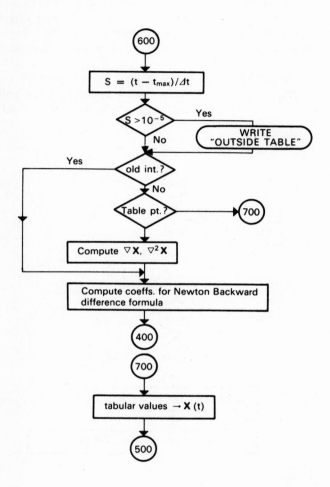

Figure 9.5.   (Continued).

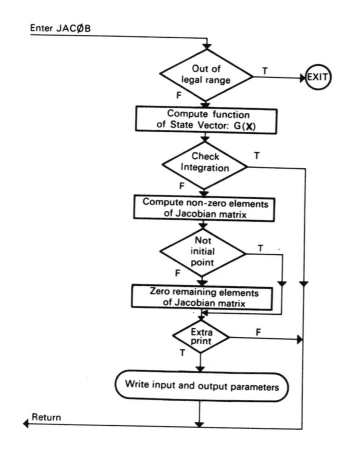

Figure 9.6.   Example:
Boundary Layer on a Wedge—Jacobian Subroutine—JACOB.

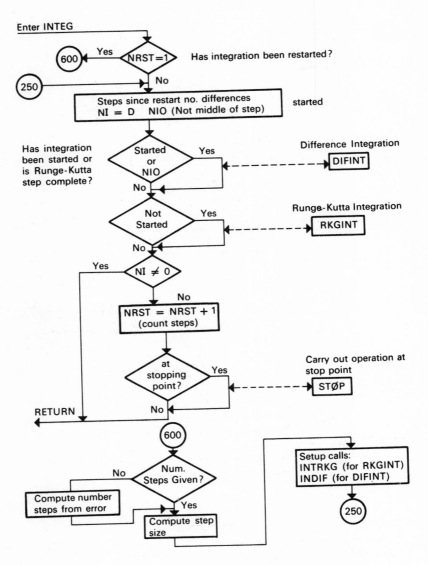

Figure 9.7. Integration Control Subroutine—INTEG.

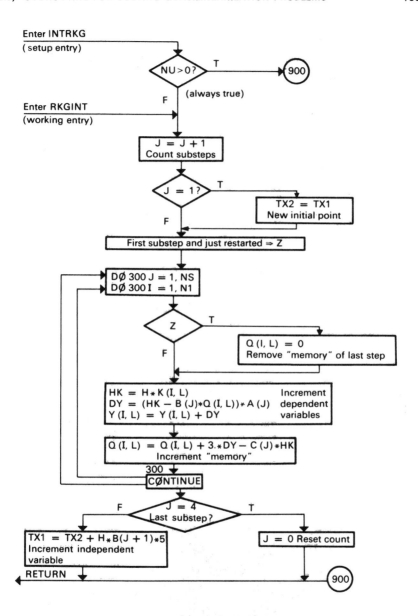

Figure 9.8. Runge-Kutta-Gill Integration Subroutine—INTRKG.

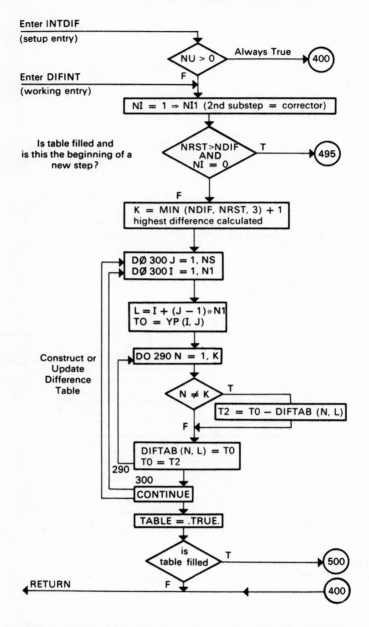

Figure 9.9. Difference Integration Subroutine—INTDIF.

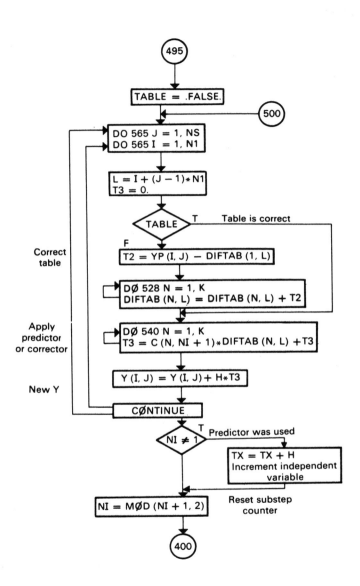

Figure 9.9.   (Continued).

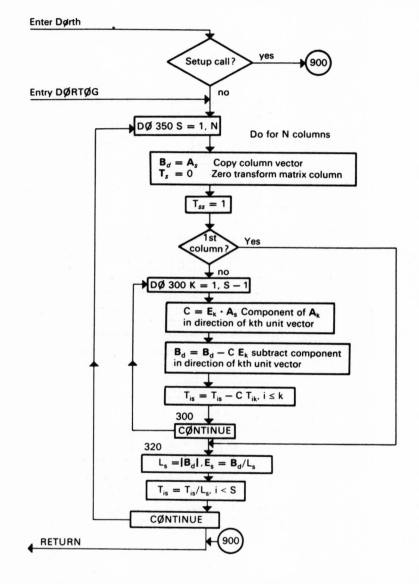

Figure 9.10.   Orthogonalization Subroutine—DORTH.

Appendix 1

PROGRAM FOR COMPUTING
BOUNDARY LAYER UPON A WEDGE

PROGRAM FOR COMPUTING
BOUNDARY LAYER UPON A WEDGE

MAIN PROGRAM

```
C     BOUNDARY LAYER ON A WEDGE, MAIN PROGRAM
C     ********************************************************************
      REAL   H1(3,1,101), RHOS(20)
      DOUBLE PRECISION XKS(3,101), P(3,101), JACM(3,3), T(1,1,1),
     A  CTP(1,1), DXKP1(3,2), XKP1(3,2), ACUM(3,2), XI, IA, TSUB, B(10),
     B  VARIND, BETA, DATE
      LOGICAL LC
      COMMON /ZWEDGE/ BETA
     A      /ZX/ XI(10), EPSP, D, IA(20)
     B      /ZSUB/ KSUB(10), TSUB(10), VARIND(101)
     C      /ZI/ PAD(12), NDIF, NBP, PAD1(42)
     D      /ZOO/ N(5), E, NSTP, INTP1, B, L(10), M(10), MI(10), RHOS
      DATA N1, N2, NU, NS, NTM / 3, 1, 3, 2, 0 /
C     ----------------------------------------------------------------
   1            CALL COUNTV
                CALL CDATEV ( DATE )
   2 READ (5,3) BETA, XI(1), RHOS(1), NSTP, NDIF, INTP1, N(3), N(4),LC,
     A  ( TSUB(I), KSUB(I), I = 1, INTP1 )
   3    FORMAT ( 2D6.3, E6.3, 5I6, L6 / 6( D6.3, I6 ) )
      WRITE(6,4) DATE,BETA,XI(1),RHOS(1),NSTP,NDIF,INTP1,N(3),N(4),
     A  ( TSUB(I), KSUB(I), I = 1, INTP1 )
   4    FORMAT ( 1H1, 5X, 42HLAMINAR INCOMPRESSIBLE BOUNDARY LAYER ON A,
     A  5HWEDGE, 3X, A8 // 5X, 'INPUT DATA' //
     D  2X, 1P2D10.3, 1P1D10.3, 5I6 // 2X, 6( 1P1D11.3, 1I4 ) )
C     ----------------------------------------------------------------
   5            B(1) = 1.D0
                M(1) = 2
                L(1) = KSUB(INTP1)
                MI(1)= 1
                IMAX = KSUB(INTP1)
                KSUB( INTP1 + 1 ) = 0
                RHOS(2) = RHOS(1)
                RHOS(3) = RHOS(1)
                NBP = 0
                N(5) = 0
C            SET UP INDEPENDANT VARIABLE
   6            CALL SETUPT
C     ----------------------------------------------------------------
C     CALL QUASILINEARIZATION SUBROUTINE
   7  CALL QASLN ( XKS, P, H1, JACM, T, CTP, NTM, N1, N2, NU, IMAX,
     A            DXKP1, XKP1, ACUM, NS )
                CALL TIMEV ( EXT )
                WRITE (6,8) EXT
   8            FORMAT ( 1H0, 5X, 21HTIME AT END OF CASE = 1F8.3 )
                IF (.NOT.LC) GO TO 2
      CALL EXIT
      STOP
C     ++++++++++++++++++++++++++++++++++++++++++++++++++++++++++++++++++
      END
```

SUBROUTINE QASLN

```
C      QUASILINEARIZATION SUBROUTINE
C      *******************************************************************
       SUBROUTINE QASLN ( XKS, P, H1, XJAC, T, CTP, NTM, N1, N2, NU,
      A                   IMAX, DXKP1, XKP1, ACUM, NS )
       LOGICAL*1 STOR
       INTEGER*2 I, J, K, L, I1, I2
       DOUBLE PRECISION ACUM, TNSTOR, F, IA, DOTD, H2, H3, P, P2, C, CS,
      A   CTP, T, E, SS, SP, DBLUNT, DBLDOT, DSQRT, MU(20), D(20),
      B   MUD2 (20), TX, XKP1, FWL, XK, DOTJX, B, XI, G, XJAC, H, XKS,
      C   DXKP1, TSUB, VARIND, RHO(20), DTPRNT
       DIMENSION XKS(N1,IMAX), P(N1,IMAX), H1(N1,N2,IMAX), XJAC(N1,NU),
      A   T(N2,N2,NTM), CTP(N2,NTM), DXKP1(NU,NS), XKP1(NU,NS),ACUM(NU,NS)
       DIMENSION B(10), H2( 6, 6), N(5), G(20),
      A MEMBER(10), LOCATE(10), TEMP( 12,12) ,    C(5), MI(10)
       COMMON /ZOO/ N, MX, XERR, EPS, B, LOCATE, MEMBER, MI, RHOS(20)
       COMMON /ZT/ KTRANS(50), E(6,6), P2(6), CS, H3(6,8)
       COMMON /ZSUB/ KSUB(10),TSUB(10),   VARIND(100)
       COMMON / ZX / XI(10), EPSP, EPSP1, IA(20)
       COMMON / ZI / TX, H ,NI,N1C,NSC, NMIN,NMAX,NRST, TNSTOR, NDIF,NBP
      A , DTPRNT , XK(20)
C      -------------------------------------------------------------------
C          INITIALIZE INTEGER CONTROL VARIABLES
                   N3 = N(3)
                   N1P1 = N1 + 1
                   NC = NU - N1
                   N(1) = N1
                   N(2) = N2
                   N5  = N(5)
                   N1C = N1
                   NSC = NS
                   IIT = 2
                   NU1 = NU
                   IF ( IABS( NBP ).LE.NC ) NU1 = NU - NBP
C      -------------------------------------------------------------------
C          SET-UP CALLS FOR MULTIPLE ENTRY SUBROUTINES
                   CALL DORTH ( H2, E, IA, T, C, N2, 6, 6, 0 )
C      -------------------------------------------------------------------
C          GENERATE INITIAL GUESSED DISTRIBUTION OF DEPENDANT VARIABLES
                   CALL XIN ( XKS, N1, IMAX        )
C      -------------------------------------------------------------------
C          DO N3 ITERATIONS OF QUASILINEARIZATION PROCESS (MAY CONVERGE 1ST)
  250  DO 950 NITER = 1, N3
                   IF ( N(4).EQ.0 ) GO TO 300
                   WRITE (6,254)
  254              FORMAT( 1H1 )
                   DO 257 J = 1, IMAX
  257              WRITE (6,258) VARIND(J), (XKS(I,J), I=1,N1)
  258              FORMAT (1X, F11.3, 1P10D10.3 )
                   IF (NC.NE.0) WRITE (6,386) (XK(I)    , I=N1P1, NU )
C      -------------------------------------------------------------------
```

```
C          SET UP INITIAL CONDITION FOR PARTICULAR AND N2 HOMOGENEOUS SOLN
  300           DO 330 K = 2, NS
                DO 320 J = 1, NU
  320             XKP1( J, K )    = 0. D0
                  M = MI(K-1)
  330             XKP1( M, K ) = XI( K-1 )
                DO 351 J = 1, N1
                  G(J) = 0.D0
                  K  =  N1  +  J
                  XKP1( J, 1 )    = XKS( J, 1 )
  351           IF ( J.LE.NC ) XKP1(K,1) = XK(K)
                  MT = 1
                  KTRANS(1) = 0
C          --------------------------------------------------------------
C          COMPUTE ALGEBRAIC INITIAL CONDITIONS
                CALL ALGBC ( XKP1, G, XJAC )
                DO 385 J = 1, N1
                  IF ( G(J).NE.0.D0 ) XKP1(J,1) = G(J)
                DO 384 K = 2, NS
                  FWL = 0.D0
                  DO 380 M = 1, NU
  380               FWL = XJAC(J,M) * XKP1(M,K) + FWL
                  IF (FWL.NE.0.D0) XKP1(J,K) = FWL
                  H1(J,K-1,1) = XKP1(J,K)
  384           CONTINUE
  385           P(J,1) = XKP1(J,1)
                IF ( N(4).LT.2 ) GO TO 390
                  DO 387 J = 1, NS
  387               WRITE (6,386) ( XKP1(I,J), I = 1, NU )
  386               FORMAT ( / ( 1X, 1P12D9.2 ))
C          --------------------------------------------------------------
C          INTEGRATE THRU UP TO 10 SEGMENTS WITH BOUNDARIES GIVEN BY TSUB
  390 DO 715 KP=1,10
      IF( KSUB(KP+1).EQ.0 ) GO TO 716
       I1=KSUB(KP)
      I2 = KSUB(KP+1)-  1
      DTPRNT = VARIND(I1+1) - VARIND(I1)
                  NRST = -1
                  NI =  -1
                  WRITE (6,393) DTPRNT, I1, KSUB(KP+1)
  393               FORMAT(/3X, 18HSTORAGE INTERVAL = F16.5, 2I10 )
                  NMIN = I1
                  NMAX = KSUB(KP+1)
      DO 715 KPRINT =  I1,I2
C        INTEGRATE THRU I2-I1 EQUALLY SPACED STORAGE INTERVALS
                  KPRINT = KPRINT
                  TX = VARIND( KPRINT )
                  TNSTOR = VARIND( KPRINT + 1 )
C          --------------------------------------------------------------
C          INTERPOLATE FOR XK IN XKS
  400 IF( NI .EQ. 0 .OR. NI .EQ. 2  )    GO TO 470
                  CALL INTER ( XKS, N1, IMAX ,MU,D,MUD2,VARIND , KPRINT )
C          --------------------------------------------------------------
```

```
C          EVALUATE DERIVATIVES AND JACOBIAN
                   IF (N(4).GT.2) WRITE (6,386) XK
   450           CALL JACOB ( TX, XK, G, XJAC, 1 )
   470           DO 490 I = 1, N1
                   DOTD = G(I)
                   DO 476 J = 1, NU1
   476               DOTD =       XJAC(I,J)  * ( XKP1(J,1) -   XK(J) ) + DOTD
                   DXKP1(I,1) =        DOTD
   490           CONTINUE
                 DO 550 M = 2, NS
                 DO 550 I = 1, N1
                   DOTJX = 0.DO
                   DO 516 J = 1, NU
   516               DOTJX = XJAC(I,J) * XKP1(J,M) + DOTJX
                   DXKP1(I,M) = DOTJX
   550           CONTINUE
C          --------------------------------------------------------------
C          INTEGRATE DXKP1
                   CALL INTEG ( XKP1,DXKP1, NU,NS ,ACUM ,TSUB(KP+1) )
                   IF (NI.NE.O ) GO TO 400
               IF( DABS( TNSTOR - TX) .GT. 1.D-6 ) GO TO 400
   586           IF ( N(4) .LT. 3 ) GO TO 625
                   WRITE (6,386) TX
                   WRITE (6,386) XKP1
C          --------------------------------------------------------------
C          SAVE PARTICULAR AND HOMOGENEOUS SOLUTIONS
   625           DO 650 I = 1, N2
                     M = MEMBER( I )
                     STOR =              KPRINT.EQ.LOCATE(I) -1
                     IF(STOR) H3(I,NS) = B(I) - XKP1(M,1)
                   DO 650 J = 1, N2
                     IF (STOR) H3(I,J) = XKP1(M,J+1)
   650 CONTINUE
                   DO 660 I = 1, N1
                     P(I,KPRINT+1) = XKP1(I,1)
               DO 660 J = 1, N2
   660             H1(I,J,KPRINT+1) = SNGL( XKP1(I,J+1) )
C          --------------------------------------------------------------
C          TEST B.C. SUBSPACE OF HOMOG. SOLN. MATRIX. IF VECTORS ARE NEARLY
C          PARALLEL, CONSTRUCT NEW ORTHO-NORMAL SET, STORE TRANS. MATRIX.
C          REDUCE PART.SOLN.TO UNIT VECTOR BY SUB.HOMOG.SOLNS.
               IF(KPRINT.EQ.1.OR.KPRINT.EQ. I2 .OR.MT.GE.NTM ) GO TO 715
                   DO 665 I = 1, N2
                     M = MEMBER(I)
                   DO 664 J = 1, N2
                     TEMP(I,J) = SNGL(XKP1(M,J+1) )
   664               H2(I,J) = XKP1(M,J+1)
   665               P2(I) = XKP1(M,1)
                     SP = 1.DO
                   DO 673 J = 1, N2
                     SS = 0.DO
                   DO 672 I = 1, N2
   672               SS = H2(I,J)**2 + SS
   673               SP = SS * SP
                   IF ( SP.EQ.0.DO ) GO TO 715
                     SCALE = 1. / SQRT( SNGL( SP ) )
                     MX = IDETRM ( I2, N2,TEMP ,SCALE)
                   IF (ABS(SCALE).GE. EPSP ) GO TO 715
```

```
                  NRST = 0
                  CALL DORTOG ( T(1,1,MT), N2 )
                  CS = 1.DO - 1.DO / DBLUNT( P2, IA, N2 )
                  DO 694 K = 1, N2
                  CTP(K,MT) = CS * DBLDOT( P2, E(1,K), N2 )
                  DO 694 I = 1, NU
   694            ACUM(I,K) = XKP1(I,K+1)
   695            CALL MATMPY (ACUM,T(1,1,MT),XKP1(1,2),NU,N2,N2,NU,N2,N2)
   696            CALL MATMPY (XKP1(1,2),CTP(1,MT),IA,NU,N2,1,NU,N2,1)
                  KTRANS(MT) = KPRINT + 1
                  MT = MT + 1
                  DO 708 I = 1, NU
                  XKP1(I,1) = XKP1(I,1) - IA(I)
                  DO 708 J = 1, NS
   708            ACUM(I,J) = 0.DO
                  IF (N(4).LE.1) GO TO 715
                  WRITE (6,386) XKP1
   715  CONTINUE
   716            MTM1 = MT - 1
                  WRITE (6,717) MTM1
   717            FORMAT( 5X, 32H NUMBER OF ORTHOGONALIZATIONS =   I5    )
C        ----------------------------------------------------------------
C        SET UP AND SOLVE LINEAR ALGEBRAIC EQS.FOR COEFS.OF HOMOG.SOLNS.
                  IF ( N5.LE.N2 ) GO TO 751
                  DO 749 K = 1, N2
                  DO 749 J = 1, NS
                     ACUM(K,J) = 0.DO
                     DO 749 I = 1, N5
   749               ACUM(K,J) = H3(I,K) * H3(I,J) + ACUM(K,J)
                  DO 750 J = 1, NS
                  DO 750 I = 1, N2
   750               H3(I,J) = ACUM(I,J)
   751            IF ( N2.EQ.1) GO TO 805
                  F = 1.DO
                  DO 753 J = 1, N2
   753            G(J) =   H3(J,NS)
                  IF( N(4) .LE.1) GO TO 754
                  WRITE(6,755) H3, G
   755            FORMAT( 6D16.6 )
   754            MX = ISIMDD ( 6, N2, 1, H3, G, F, IA )
                  DO 752 I = 1, N2
   752               IA( I ) = H3 (I,1)
                  GO TO ( 850, 770, 790 ), MX
   770            WRITE (6,775)
   775            FORMAT ( 1H1, 10X, 27HOVER-FLOW OCCURED IN ISIMDD )
                  GO TO 850
   790            WRITE (6,795)
   795            FORMAT ( 1H1, 10X, 19HMATRIX WAS SINGULAR                    )
                  GO TO 980
   805            IF ( H3(1,1)   .EQ. 0.DO ) GO TO 790
                  IA(1) = H3(1,NS) / H3(1,1)
   850            WRITE (6,851)
   851            FORMAT ( 20HO COMBINATION COEFF    )
                  WRITE (6,386) ( IA(I), I = 1, N2 )
C        ----------------------------------------------------------------
```

```
C          COMBINE P AND H SOLNS TO OBTAIN X(K+1)
                   DO 852 I = 1, NU
      852          RHO(I) = 0.DO
                   DO 840 I = 1, NC
                   M = I + N1
                   DO 835 K = 1, N2
      835          XKP1(M,1) = IA(K) * XKP1(M,K+1)+XKP1(M,1)
                   RHO(M) = DABS( XKP1(M,1)    - XK(M) )
      840          XK(M) = XKP1(M,1)
                   DO 895 J = 1, IMAX
                   L = IMAX + 1 - J
                   DO 872 I = 1, N1
                   DOTD = P(I,L)
                   DO 866 K = 1, N2
      866             DOTD = IA(K) * H1(I,K,L) + DOTD
                   RHO(I) = DMAX1 ( DABS( DOTD - XKS(I,L) ), RHO(I) )
      872          XKS(I,L) = DOTD
C                  TRANSFORM COMBINATION COEF VECTOR IF SOLN WAS TRANS
                   IF( L-1 .NE. KTRANS(MT-1) .OR. MT.LE.1 ) GO TO 895
                   MT = MT - 1
                   DO 886 I = 1, N2
      886          C(I) = IA(I) - CTP(I,MT)
                   CALL MATMPY ( T(1,1,MT), C, IA, N2, N2, 1,N2, N2, 1 )
      895          CONTINUE
C          ----------------------------------------------------------------
                   CALL TIMEV ( EXT )
                   WRITE (6,898) EXT
      898             FORMAT ( 1H0, 10X, 1F7.3,1X, 7HSECONDS )
                   WRITE (6,910)
      910             FORMAT ( 31H0 CONVERGENCE PARAMETERS----RHO   )
                   WRITE (6,386) ( RHO(I), I = 1, NU )
                   DO 945 I = 1, NU
                   RHOSS = RHO(I)
                   IF ( RHOSS .GE. RHOS(I) ) GO TO 949
      945          CONTINUE
                   WRITE (6,946) NITER
      946             FORMAT ( / 9X, 29H PROCESS CONVERGED, ITER.NO.= I5 )
                   GO TO 980
      949          CALL OUTPUT ( XKS, N1, IMAX, 1 )
      950 CONTINUE
C          ----------------------------------------------------------------
                   IIT = 3
                   WRITE (6,965) N3
      965             FORMAT ( 1H1, 10X, 17HNO CONVERGENCE IN,I3,6HCYCLES )
      980 CALL OUTPUT( XKS,N1,IMAX,IIT)
          IF( IIT .EQ. 0 ) GO TO 1110
C          ----------------------------------------------------------------
```

```
C        CHECK SOLUTION BY SOLVING INITIAL VALUE PROB FROM CONVERGED CONDS.
         WRITE(6,981)
     981 FORMAT(54H1 SOLUTION CHECK BY INTEGRATION OF INITIAL CONDITIONS   )
                 DO 1000 I = 1, N1
                    RHO(I) = 0.D0
                    ACUM(I,1) = 0.D0
    1000            XKP1(I,1) = XKS(I,1)
                 DO 1001 I = N1P1, NU
                    XKP1(I,1) = XK(I)
    1001            ACUM(I,1) = 0.D0
C        ----------------------------------------------------------------
         DO 1050 I = 1, 10
                 I1 = KSUB(I)
                 I2 = KSUB(I+1) - 1
                 DTPRNT = VARIND(I1+1) - VARIND(I1)
                 NRST = -1
    1020         DO 1050 J = I1, I2
                    TX = VARIND(J)
                    TNSTOR = VARIND(J+1)
    1030          CALL JACOB ( TX, XKP1, DXKP1, XJAC, 2 )
                 CALL INTEG ( XKP1, DXKP1, N1, 1, ACUM, TSUB(I+1) )
                 IF ( NI.NE.0 ) GO TO 1030
                 IF ( DABS( TNSTOR - TX ) .GT. 1.D-6 ) GO TO 1030
                 DO 1040 K = 1, N1
    1040             RHO(K) = DMAX1( DABS( XKS(K,J+1) - XKP1(K,1)), RHO(K) )
                 IF ( J .EQ. IMAX-1 ) GO TO 1100
    1050 CONTINUE
    1100     WRITE (6,910)
            WRITE (6,386) ( RHO(I), I = 1, N1 )
    1110 CALL OUTPUT( XKS,N1 , IMAX, 4 )
     990 RETURN
C        ----------------------------------------------------------------
         END
```

SUBROUTINE XIN


```
C      INITIAL APPROXIMATION TO STATE VECTOR AS FUNCT OF IND VARIABLE
C      ***********************************************************************
       SUBROUTINE XIN ( Y, NU, IMAX )
       LOGICAL FOWARD
       DOUBLE PRECISION Y, Z, W, XK, TSUB, VARIND, X
       COMMON /ZT/ M(15), L(17), Z(50), W(50)
       COMMON /ZSUB/ KSUB(10), TSUB(10), VARIND(100)
       COMMON /ZI/ PAD(16), XK(20)
       DIMENSION Y(NU,IMAX) ,MC(15)
C      ------------------------------------------------------------------
       READ (5,160)N1, NC, M
  160    FORMAT(18I4)
           IF (NC.GE.0) GO TO 240
C              READ PARAMETERS, STORE IN STATE VECTOR, XK
               NC = - NC
  190              READ (5,200) (W(K),K=1,NC)
  200                FORMAT ( 12D6.3)
                    DO 220 K=1,NC
                       N1PK  = N1 + K
  220              XK(N1PK) = W(K)
                   GO TO 900
C      ------------------------------------------------------------------
  240    IF (N1) 250,900,280
  250        N1 = -N1
             FOWARD = .FALSE.
           GO TO 290
C      ------------------------------------------------------------------
  280          FOWARD = .TRUE.
  290      L(1) = 1
           DO 310 I =1, N1
           MC(I) = 1
  310      L(I+1) = M(I) + L(I)
           LT = L(N1+1) - 1
           READ (5,200) (Z(K), W(K), K = 1, LT)
             DO  540 N = 1, IMAX
               IF (FOWARD) GO TO 390
                 J = IMAX + 1 - N
                 X = VARIND(IMAX) - VARIND(J)
                 GO TO 410
C      ------------------------------------------------------------------
  390            X = VARIND(N)
                 J = N
  410            DO 540 I = 1, N1
                   K = L(I)
                   IF ( M(I).NE.1 ) GO TO 470
                   Y(I,J) = W(K)
                   GO TO 540
C      ------------------------------------------------------------------
```

```
      460              MC(I) = MC(I) + 1
                      IF (MC(I).GT.M(I)) GO TO 510
                         L(I) = L(I) + 1
      465              K = L(I)
      470              IF ( X - Z(K+1))510, 475, 460
      475                 Y(I,J) = W(K+1)
                      GO TO 540
C     ----------------------------------------------------------------
      510                 K1 = K + 1
                      Y(I,J) = W(K) + (X-Z(K)) * (W(K1)-W(K)) / (Z(K1)-Z(K))
      540     CONTINUE
              IF ( NC.EQ.0) GO TO 900
              GO TO 190
      900 RETURN
C     ----------------------------------------------------------------
          END


                        SUBROUTINE INTER


C     2ND ORDER BESSEL INTERPOLATION FOR QASLIN
C     REF.INTRODUCTION TO NUMERICAL ANAYSIS,HILDEBRAND,MCGRAW-HILL,CH4
C     ****************************************************************
      SUBROUTINE INTER ( X, N1, IMD, MU, D, MUD2, VARIND, II )
      DOUBLE PRECISION X(N1,IMD), MU(N1), D(N1), MUD2(N1), VARIND(IMD),
     A  C1, C2, S, TX, XK, H, DT
      COMMON /ZI/ TX, H, NI, NV, NS, IN, IM, PAD(5), DT, XK(12)
      DATA IMIN0, I0 / 0, -1000 /
C     ----------------------------------------------------------------
C        COMPUTE INTERVAL NUMBER AND WHETHER BELOW, IN, OR ABOVE TABLE
                 I = II + IDINT( ( TX - VARIND(II) + 1.D-4 )/ DT )
                 IF (I-IN) 200, 250, 300
C     ----------------------------------------------------------------
C        TX IS IN 1ST INTERVAL OR BELOW TABLE, USE NEWTON FORWARD DIFF.
      200        WRITE (6,210) TX
      210        FORMAT (/ 5X, 9HARGUMENT=F15.5, 17H IS OUTSIDE TABLE )
      250        S = ( TX - VARIND(IN) ) / DT
               IF ( I.EQ.I0 .AND. IN.EQ.IMIN0 ) GO TO 292
               IMIN0 = IN
               IF ( DABS(S) .LE. 1.D-5 ) GO TO 700
               I0 = I
               DO 290 J = 1, N1
                 MU(J)   = X(J,IN)
                 D(J)    = X(J,IN+1) - X(J,IN)
      290        MUD2(J) = X(J,IN+2) - 2.D0*X(J,IN+1) + X(J,IN)
      292        C1 = S
               C2 = S * ( S - 1.D0 ) * .5D0
               GO TO 400
C     ----------------------------------------------------------------
      300        IF ( I.GE.IM-1 ) GO TO 600
```

```
C          TX IS IN INTERIOR INTERVAL, USE BESSEL INTERPOLATION
   310            S = ( TX - VARIND(I) ) / DT
                  IF ( I.EQ.IO ) GO TO 380
                  IF ( DABS(S).LE.1.D-5 ) GO TO 700
                  IO = I
                  DO 360 J = 1, N1
                  MU(J)   = .5D0 * ( X(J,I) + X(J,I+1) )
                  D(J)    = X(J,I+1) - X(J,I)
   360            MUD2(J) = .5D0 * ( X(J,I-1) + X(J,I+1) ) - MU(J)
   380            C1 = S - .5D0
                  C2 = S * ( S - 1.D0 ) * .5D0
   400            DO 410 J = 1, N1
   410            XK(J) = MU(J) + C1*D(J) + C2*MUD2(J)
   500 RETURN
C     --------------------------------------------------------------------
C          TX IS IN LAST INTERVAL OR BEYOND TABLE,USE NEWTON BACKWARD DIFF.
   600            S = ( TX - VARIND(IM) ) / DT
                  IF ( S.GT.1.D-5 ) WRITE (6,210) TX
                  IF ( I.EQ.IO ) GO TO 650
                  IF ( DABS(C2) .LE. 1.D-5 ) GO TO 700
                  IO = I
                  DO 640 J = 1, N1
                  MU(J)   = X(J,IM)
                  D(J)    = X(J,IM) - X(J,IM-1)
   640            MUD2(J) = X(J,IM) - 2.D0*X(J,IM-1) + X(J,IM-2)
   650            C1 = S
                  C2 = ( S + 1.D0 ) * S * .5D0
                  GO TO 400
C     --------------------------------------------------------------------
C          TX IS AT A TABULAR POINT
   700            DO 720 J = 1, N1
   720            XK(J) = X(J,I)
                  GO TO 500
C     ++++++++++++++++++++++++++++++++++++++++++++++++++++++++++++++++++++
      END
```

SUBROUTINE ALGBC

```
C     BOUNDARY LAYER ON A WEDGE, DUMMY ALGEBRAIC INITIAL BOUNDARY COND.
C     ********************************************************************
      SUBROUTINE ALGBC ( X, G, J )
      DOUBLE PRECISION X, G(3), J(3,3)
C     --------------------------------------------------------------------
                  DO 2 K = 1, 3
                  DO 1 I = 1, 3
   1              J(I,K) = 0.D0
   2              G(K)   = 0.D0
      RETURN
C     ++++++++++++++++++++++++++++++++++++++++++++++++++++++++++++++++++++
      END
```

SUBROUTINE WEDGJ

```
C     BOUNDARY LAYER ON A WEDGE, JACOBIAN MATRIX
C     *********************************************************************
      SUBROUTINE JACOB ( ETA, X, G, J, ISW )
      DOUBLE PRECISION ETA, X(3), G(3), J(3,3), F, FP, F2P, BETA, T, H
      COMMON /ZWEDGE/ BETA
    A      /ZI/      T, H, NI, PAD(3), NRST, PAD1(46)
    B      /ZOO  / N(5), PAD2(81)
C     ------------------------------------------------------------------
            IF ( ETA.GT.9.DO ) CALL EXIT
    1           F2P = X(1)
                FP  = X(2)
                F   = X(3)
C       G(I) = DX(I)/DETA
                G(1) =(-F*F2P + BETA*( FP**2 - 1.DO )) * .5DO
                G(2) = F2P
                G(3) = FP
            IF (ISW.EQ.2) GO TO 9
C     ------------------------------------------------------------------
C       JACOBIAN MATRIX J(I,J) = DG(I)/DX(J)
                J(1,1) = -F * .5DO
                J(1,2) = FP * BETA
                J(1,3) = -F2P * .5DO
                J(2,1) = 1.DO
                J(3,2) = 1.DO
            IF (ETA.NE.0.DO) GO TO 7
                J(2,2) = 0.DO
                J(2,3) = 0.DO
                J(1,3) = 0.DO
                J(3,3) = 0.DO
    7           IF ( N(4).LE.2 ) GO TO 9
    8           WRITE (6,2) ETA, H, NI, NRST
    2             FORMAT ( / 5X, 1P2D12.3, 2I6 )
                WRITE (6,3) X, G, J
    3             FORMAT ( 5X, 1P3D12.3 )
    9 RETURN
C     ++++++++++++++++++++++++++++++++++++++++++++++++++++++++++++++++++++
      END
```

SUBROUTINE OUTPUT

```
C     BOUNDARY LAYER ON A WEDGE, OUTPUT SUBROUTINE
C     *************************************************************************
      SUBROUTINE OUTPUT ( X, N1, IMAX, IIT )
      DOUBLE PRECISION X(N1,IMAX), TSUB, VARIND, BETA
      COMMON /ZSUB/    KSUB(10), TSUB(10), VARIND(101)
    A         /ZWEDGE/ BETA
C     -------------------------------------------------------------------
    1       IF ( IIT.EQ.1 .OR. IIT.EQ.4 ) GO TO 9
              WRITE (6,2) BETA, ( VARIND(I),( X(J,I),J=1,3),I=1,IMAX)
    2   FORMAT ( 1H1, 10X, 25HBOUNDARY LAYER ON A WEDGE // 5X, 6HBETA =
    A   F7.3 // 4X, 3HETA, 5X, 3HF'', 8X, 2HF', 9X, 1HF /( 2X, OPF7.3,
    B   1P3D12.4 ) )
    9 RETURN
C     +++++++++++++++++++++++++++++++++++++++++++++++++++++++++++++++++++++++++
      END
```

SUBROUTINE STOP

```
C     STOP SUBROUTINE FOR QASLN (DUMMY)
C     *************************************************************************
      SUBROUTINE STOP ( X, DX, NU, NS, ISW )
      RETURN
C     +++++++++++++++++++++++++++++++++++++++++++++++++++++++++++++++++++++++++
      END
```

SUBROUTINE DBLDOT

```
C     DOUBLE PRECISION DOT PRODUCT
C     *************************************************************************
      REAL FUNCTION DBLDOT*8 (A, B, N )
      REAL*8 DBLDOT, A(N), B(N)
      INTEGER*2 I
C     -------------------------------------------------------------------
    6           DBLDOT = 0.D0
    7       DO 1 I = 1, N
    1           DBLDOT = A(I) * B(I) + DBLDOT
    4   RETURN
C     -------------------------------------------------------------------
      END
```

SUBROUTINE DBLUNT

```
C       DOUBLE PRECISION UNIT VECTOR FUNCTION
C       ***********************************************************************
        REAL FUNCTION DBLUNT*8 ( A, UNIT, N )
        REAL*8 A(N),          UNIT(N), DBLUNT, DBLDOT, MAGI
        INTEGER*2 I
C       -------------------------------------------------------------------
    3              DBLUNT = DSQRT( DBLDOT(A,A,N) )
                   MAGI = 1.D0 / DBLUNT
               DO 5 I = 1, N
    5              UNIT(I) = A(I) * MAGI
        RETURN
C       ++++++++++++++++++++++++++++++++++++++++++++++++++++++++++++++++++++
        END
```

SUBROUTINE MATMPY

```
C       DOUBLE PRECISION MATRIX MULTIPLY
C       ***********************************************************************
        SUBROUTINE MATMPY ( A, B, C, N1, N2, N3, ND1, ND2, ND3 )
        REAL*8 A(ND1,ND2), B(ND2,ND3), C(ND1,ND3), SUM
C       -------------------------------------------------------------------
        DO 2 K = 1, N3
            DO 2 I = 1, N1
                SUM = 0.D0
                DO 1 J = 1, N2
    1               SUM = A(I,J) * B(J,K) + SUM
    2           C(I,K) = SUM
        RETURN
C       ++++++++++++++++++++++++++++++++++++++++++++++++++++++++++++++++++++
        END
```

SUBROUTINE DORTH

```
C     SCHMIDT ORTHOGONALIZATION PROCEDURE TO CONSTRUCT THE MATRIX B.
C     (REF.METHODS OF APPLIED MATHEMATICS, F.B.HILDEBRAND, PR-H1952,P34)
C      E=MATRIX UNIT VECTORS OF B, L=LENGTHS OF VECTORS IN B , E = AT
C     **************************************************************
      SUBROUTINE DORTH ( A, E, L, T, BD, N, NA, NE, NT )
      INTEGER S
      DOUBLE PRECISION A(NA, N), E(NE, N),  T(NT,NT), L(N), BD(N), C,
     A               DBLDOT, DBLUNT
C     ------------------------------------------------------------
C         IF SET UP CALL, RETURN IMMEDIATELY
          IF (NT.EQ.0) GO TO 900
   70         CONTINUE
      ENTRY DORTOG ( T, NT )
  100     DO 350 S = 1, N
              DO 170 I = 1, N
                  BD(I) = A(I,S)
  170             T(I,S) = 0.D0
                  T (S,S) = 1.D0
              IF (S.EQ.1) GO TO 320
  195             M = S - 1
              DO 300 K = 1, M
                  C = DBLDOT ( E(1,K), A(1,S), N )
                  DO 300 I = 1, N
                  BD(I) = BD(I) - C * E (I,K)
                  IF (I.LT.S.AND.I.LE.K) T(I,S) = T(I,S) - C*T(I,K)
  300         CONTINUE
  320             L(S) = DBLUNT( BD, E(1,S), N )
                  DO 350 I = 1, N
                  IF (I.LE.S) T(I,S) = T(I,S) / L(S)
  350     CONTINUE
C     ------------------------------------------------------------
  900 RETURN
C     ------------------------------------------------------------
      END
```

SUBROUTINE INTEG

```
C     SUBROUTINE CONTROLS INTEGRATION
C     ***************************************************************
      SUBROUTINE INTEG(X,DX,NU,NS,ACUM,TSTOP )
      LOGICAL STARTD, NIO
      DOUBLE PRECISION H, T, TSTOP, X, DX, ACUM, DT
      EQUIVALENCE (XERR,NSTEP)
      DIMENSION   X(NU,NS),DX(NU,NS),ACUM(NU,NS)
      COMMON /ZOO/ PAD1(6),XERR,PAD2(79)
      COMMON /ZI/ T, H, NI, PAD3(4), NRST, PAD4(2), NDIF, PAD5, DT,O(40)
C     ------------------------------------------------------------------
      IF( NRST .EQ. -1 ) GO TO 600
C         INTEGRATE USING DIFFERENCES OR BUILD TABLE IF FIRST NDIF STEPS
  250             STARTD = NRST.GE.NDIF
                  NIO    = NI.EQ.O
              IF ( STARTD .OR. NIO ) CALL DIFINT
C                 USE RUNGE-KUTTA-GILL TO START INTEGRATION
              IF ( .NOT. STARTD )    CALL RKGINT
              IF ( NI.NE.O   ) GO TO 500
              NRST = NRST & 1
              IF ( DABS( T-TSTOP ).LE.TOL ) CALL STOP( X, DX, NU, NS, 2)
  500 RETURN
C     ------------------------------------------------------------------
C     RESTART BRANCH
  600             NRST = 0
                  NI = 0
C     ------------------------------------------------------------------
C         COMPUTE STEP SIZE
  650         IF(IABS(NSTEP).LE.100000) GO TO 656
              KSTEPM = SNGL(DT) / XERR ** .2
              GO TO 660
  656             KSTEPM = NSTEP
  660     IF(IABS(KSTEPM).GT.0) GO TO 690
              KSTEPM = 1
  690             H = DT/DBLE( FLOAT( KSTEPM ) )
                  TOL = .1D0 * H
C     ------------------------------------------------------------------
C         INITIALIZE RUNGE-KUTTA AND DIFFERENCE SUBROUTINES
              CALL INTRKG ( X, DX, NU, NS, ACUM )
              CALL INTDIF ( X, DX, NU, NS, ACUM )
              GO TO 250
C     &&&&&&&&&&&&&&&&&&&&&&&&&&&&&&&&&&&&&&&&&&&&&&&&&&&&&&&&&&&&&&&&&&&&&
      END
```

SUBROUTINE SETUP

```
C     SUBROUTINE TO SET UP INDEPENDENT VARIABLE FOR QUASILINEARIZATION
C     *****************************************************************
      SUBROUTINE SETUPT
      DOUBLE PRECISION TSUB, VARIND, DT
      COMMON /ZSUB/ KSUB(10), TSUB(10),VARIND(100)
      COMMON /ZOO/ PAD1( 7), NINTP1, PAD2(78)
C     -----------------------------------------------------------------
100       DO 500 I = 2, NINTP1
              IM1 = I - 1
              NDK = KSUB(I) - KSUB(IM1)
              DK = NDK
              DT = ( TSUB(I) - TSUB(IM1) ) / DBLE( DK )
              KSIM1 = KSUB(IM1)
              VARIND( KSIM1 ) = TSUB( IM1 )
              NDKM1 = NDK - 1
200           DO 400 J = 1, NDKM1
              K = KSUB(IM1) + J
              AJ = J
              VARIND(K) = DT * DBLE( AJ ) + TSUB( IM1 )
400           CONTINUE
500       CONTINUE
              VARIND(K+1) = TSUB( NINTP1 )
      RETURN
C     -----------------------------------------------------------------
      END
```

SUBROUTINE INTRKG

```
C     RUNGE-KUTTA-GILL 4TH ORDER INTEGRATION SUBROUTINE FOR QASLIN
C     REF MATHEMATICAL METHODS FOR DIGITAL COMPUTERS, CHAP.9
C     *****************************************************************
      SUBROUTINE INTRKG (Y,K,NU,NS,Q)
      LOGICAL Z
      DOUBLE PRECISION H , HK, DY, Y, A(4), B(4), C(4), Q,K,TX1,TX2
      DIMENSION Y(NU,NS),K(NU,NS),Q(NU,NS)
      COMMON /ZI/ TX1, H, J, N1, NH, PAD1(2), NRST, PAD2(46)
      DATA A / .5D0, .2928932D0, 1.7071068D0, .1666666666666667D0 /,
     *     B / 2.D0, 2*1.D0, 2.D0 /,
     *     C / .5D0, .2928932D0, 1.7071068D0, .5D0 /
C     -----------------------------------------------------------------
C     SET UP ENTRY
              IF (NU.GT.0) GO TO 900
45            CONTINUE
C     -----------------------------------------------------------------
```

```
      ENTRY RKGINT
   50           J = J + 1
                IF(J.EQ.1) TX2 = TX1
                Z = J.EQ.1 .AND. NRST.LE.0
  200      DO 300 L = 1, NS
           DO 300 I = 1, N1
                IF ( Z ) Q(I,L) = 0.D0
                HK = H  *        K(I,L)
                DY =(HK - B(J) * Q(I,L)) * A(J)
                Y(I,L) = Y(I,L) + DY
                Q(I,L) = Q(I,L) + 3.D0*DY - C(J) * HK
  300      CONTINUE
      ------------------------------------------------------------------
           IF ( J.EQ.4 ) GO TO 350
                TX1 = TX2 +        H  * B(J+1) * .5D0
           GO TO 900
  350           J = 0
  900 RETURN
      ------------------------------------------------------------------
      END


                       SUBROUTINE INTDIF


C     DIFFERENCE INTEGRATION SUBROUTINE, ADAMS-MOULTON, FOR QASLIN
C     REF. INTRODUCTION TO NUMERICAL ANALYSIS, F.B.HILDERAND, P.199
C     *************************************************************
      SUBROUTINE INTDIF (Y,YP,NU,NS,TS)
      LOGICAL NI1, TABLE
      DOUBLE PRECISION Y,YP,TS,T0,T2,C(4,2),G(4),H,DIFTAB,TX,TNSTOR,T3
      DIMENSION Y(NU,NS), YP(NU,NS), TS(NU,NS)
      COMMON /ZI/ TX, H, NI, N1, NH, P1(2), NRST, TNSTOR, NDIF, P(43)
      COMMON /ZDIF/ DIFTAB(4,72)
      DATA C / 1.D0, .5D0, .4166666666666667D0, .375D0, 1.D0, -.5D0,
     *    -.8333333333333D-1, -.4166666666666667D-1/, G / 2.D0, 6.D0,
     *    1.D+1, 1.4D+1 /
C     ------------------------------------------------------------------
C     SET UP ENTRY
           IF (NU.GT.0) GO TO 400
   94           CONTINUE
C     ------------------------------------------------------------------
```

```
      ENTRY DIFINT
  97              NI1 = NI .EQ. 1
      IF ( NRST.GT.NDIF .AND. NI.EQ.0 ) GO TO 495
         CONSTRUCT OR UPDATE DIFFERENCE TABLE
 200              NT = NI * NS
                  K  = MINO( NDIF, NRST, 3 ) + 1
            DO 300 J = 1, NS
            DO 300 I = 1, NI
                  L = I + (J-1) * NI
                  TO      = YP(I,J)
                  DO 290 N = 1, K
                     IF (N.NE.K) T2 = TO     - DIFTAB(N,L)
                     DIFTAB(N,L) = TO
 290                 TO      = T2
 300           CONTINUE
                  TABLE = .TRUE.
 350              IF ( NRST .GE. NDIF ) GO TO 500
 400 RETURN
C     ----------------------------------------------------------------
C     PREDICT Y USING OPEN TYPE INTEGRATION FORMULA
C     CORRECT Y USING CLOSED TYPE INTEGRATION FORMULA
 495              TABLE = .FALSE.
 500         DO 565 J = 1, NS
             DO 565 I = 1, NI
                  L = I + (J-1) * NI
                  T3 = 0.DO
             IF ( TABLE ) GO TO 530
                  T2 = YP(I,J) - DIFTAB(1,L)
                  DO 528 N = 1, K
 528              DIFTAB(N,L) = DIFTAB(N,L) + T2
 530              DO 540 N = 1, K
 540                 T3 = C(N,NI+1) *          DIFTAB(N,L)    + T3
                  IF ( .NOT. NI1 ) TS(I,J) = 0.DO
                  TS(I,J) = H * T3 - TS (I,J)
                  Y(I,J)  = Y(I,J) + TS(I,J)
                  IF (NI1.AND.Y(I,J).NE.0.DO) TS(I,J)=TS(I,J)/(G(K)*Y(I,J))
 565         CONTINUE
                  IF ( .NOT.NI1 ) TX = TX + H
                  NI = MOD( NI+1, 2 )
                  GO TO 400
C     ----------------------------------------------------------------
      END
```

AUTHOR INDEX

Numbers in parentheses indicate the numbers of the references when these are cited in the text without the name of the author.

Numbers set in *italics* designate those page numbers on which the complete literature citations are given.

223

SUBJECT INDEX

Date Due

JUN 29 1973 CIRC			